Terrorism in America

Terrorism in America

Edited by

Kevin Borgeson, PhD
Salem State College
Salem, MA

Robin Valeri, PhD
St. Bonaventure University
St. Bonaventure, NY

JONES AND BARTLETT PUBLISHERS
Sudbury, Massachusetts
BOSTON TORONTO LONDON SINGAPORE

World Headquarters
Jones and Bartlett Publishers
40 Tall Pine Drive
Sudbury, MA 01776
978-443-5000
info@jbpub.com
www.jbpub.com

Jones and Bartlett Publishers
Canada
6339 Ormindale Way
Mississauga, ON L5V 1J2
Canada

Jones and Bartlett Publishers
International
Barb House, Barb Mews
London W6 7PA
United Kingdom

Jones and Bartlett's books and products are available through most bookstores and online booksellers. To contact Jones and Bartlett Publishers directly, call 800-832-0034, fax 978-443-8000, or visit our website www.jbpub.com.

Substantial discounts on bulk quantities of Jones and Bartlett's publications are available to corporations, professional associations, and other qualified organizations. For details and specific discount information, contact the special sales department at Jones and Bartlett via the above contact information or send an email to specialsales@jbpub.com.

Production Credits
Acquisitions Editor: Jeremy Spiegel
Editorial Assistant: Maro Asadoorian
Editorial Assistant: Catie Heverling
Production Manager: Julie Champagne Bolduc
Production Assistant: Jessica Steele Newfell
Associate Marketing Manager: Lisa Gordon
Manufacturing and Inventory Control Supervisor: Amy Bacus
Composition: Lynn L'Heureux
Cover Design: Kristin E. Ohlin
Cover Image: © Digital Vision/age fotostock
Printing and Binding: Malloy, Inc.
Cover Printing: Malloy, Inc.

Library of Congress Cataloging-in-Publication Data
Terrorism in America / [edited by] Kevin Borgeson, Robin Valeri.
p. cm.
Includes index.
ISBN 978-0-7637-5524-9 (pbk. : alk. paper)
1. Terrorism--United States. I. Borgeson, Kevin. II. Valeri, Robin.
HV6432.T499 2009
363.3250973--dc22

2008028934

6048
Printed in the United States of America
12 11 10 10 9 8 7 6 5 4 3 2

This book is dedicated to Pam, Jade, and Liam. Without their support and feedback, this project never would have got off the ground. I would also like to dedicate this book to Michael E. Brown, an incredible mentor and friend.

— Kevin Borgeson

This book is dedicated to my husband, Raymond P. Valeri. Thank you for your love and support.

— Robin Maria Valeri

Table of Contents

Preface

The purpose of this volume is to provide an understanding of domestic terrorism in the United States. It is not surprising that the tragic events of September 11, 2001, focused the attention of this nation on international sources of terrorism. What is surprising is that after the April 19, 1995, bombing of the Murrah Federal Building in downtown Oklahoma City by Timothy McVeigh, an American, the public's fear and attention were focused on international sources of terrorism. In fact, antiterrorism laws rushed through Congress after the bombing of the Murrah Federal Building targeted foreign sources of terrorism. This raises two important questions: Why are Americans reluctant to look within their borders for sources of terrorism? And, what are the consequences of this oversight?

To address these questions, we begin by exploring the major sources of terrorism in the United States, comparing and contrasting these domestic sources to those in other countries. Based on data gathered from interviews, we present an image of the people who join these groups, their motivations for joining, and the fundamental beliefs shared by many of these groups. Although the focus of this volume is on domestic terrorism and the importance of understanding and preventing it, efforts by domestic terrorists to form alliances with foreign groups also are discussed. In addition, this volume includes a history of counterterrorism in the United States, with a focus on present-day efforts to combat terrorism. Lastly, research regarding the fear of terrorism, including its impact on individuals and on the nation as a whole, is presented.

As part of this volume, several theories of terrorism are discussed. However, social psychological theories are used to explain many of the beliefs and behaviors of these groups as well as the public's response to terrorism.

It also is important to note that the reader will encounter the ideologies and views of many domestic terrorist organizations within this book. It is our belief that to firmly understand and grasp the rationalizations of these groups, their views—however offensive—must be demonstrated and paraphrased in great detail. The level of depth and clarity in which these views are presented are meant solely for academic purposes and do not represent the personal beliefs of the authors or the publisher.

Lastly, we set out to produce a volume that is academically rigorous, grounded in research and theory, as well as accessible to the general public. It is our hope that this book will raise the public's awareness and concern about domestic terrorism, foster a growing body of research about these groups and their links to international terrorism, and stimulate efforts to curtail their actions.

— Robin Valeri & Kevin Borgeson

Acknowledgments

We express our deepest thanks to the contributing authors for their hard work, attention to detail, and adherence to strict deadlines.

We extend our gratitude to our editor, Jeremy Spiegel, for his patience, support, and encouragement throughout this process. We also extend our gratitude and respect to the staff at Jones and Bartlett Publishers who helped bring this volume to completion.

Acknowledgments

Contributors

Kevin Borgeson is Assistant Professor in the Criminal Justice Department at Salem State College, Salem, MA, where he teaches courses in crime scene investigation, profiling, and bias crimes. Borgeson's work has appeared in *Journal of Applied Sociology, Michigan Sociological Review*, and *American Behavioral Science.*

Stephen E. Costanza is Assistant Professor of Criminology and Criminal Justice at Central Connecticut State University. He received his PhD in Sociology from Louisiana State University. Costanza's specialty areas are spatial analysis and policy analysis. He has been published in *Criminal Justice Review, Journal of Crime and Justice, Policing and Society*, and *Journal of Social and Ecological Boundaries.*

Ronald Helms earned a BA at California State University, Chico, and MS and PhD degrees at the University of Oregon. He is currently a member of the Sociology faculty at Western Washington University where he teaches courses in criminology, policing, local jails, research methods, and statistics. In his research, Helms has used aggregate time series methods or generalized linear modeling techniques to investigate determinants of various punishment/social control outcomes. His research has appeared in *American Journal of Sociology, Social Forces, Police Quarterly, Policing and Society*, and *Social Science Research.*

John C. Kilburn, Jr., PhD, is the Department Chair of Behavioral, Applied Sciences, and Criminal Justice at Texas A&M International University in Laredo. His work has been published in journals such as *Criminal Justice Review, Urban Affairs Review*, and *Social Forces.*

Ashley Marie Nellis, PhD, earned her doctorate in 2007 from American University. Using original data collected from residents of New York City and Washington, DC, her dissertation study examined relationships among media exposure, terrorism fear, perceived risk of terrorism, and antiterrorism policy preferences. She worked closely with the University of Maryland's National Consortium for the Study of Terrorism and Responses to Terrorism (START) as a research fellow from 2005 to 2007 and has presented her findings at numerous academic conferences.

Thomas R. O'Connor teaches in and manages the undergraduate program in Criminal Justice/Homeland Security at Austin Peay State University in Clarksville, Tennessee, where he also directs the Institute for Global Security Studies. O'Connor has written a number of articles on terrorism and security topics and is the author of the Internet Web site known as MegaLinks in Criminal Justice.

Robin Maria Valeri, PhD, earned her BA in Psychology and Economics from Cornell University, and her MA and PhD in Psychology from Syracuse University. She currently is a Professor of Psychology at St. Bonaventure University. Valeri's work has appeared in various psychology and sociology journals, including *Current Psychology*, *Journal of Applied Social Psychology*, *Journal of Applied Sociology*, *Michigan Sociological Review*, and *Society and Animals*.

CHAPTER 1

Domestic Terrorism

Kevin Borgeson and Robin Valeri

PAST AND PRESENT TYPOLOGIES

Brent Smith (1994), in his book *Terrorism in America*, contrasts the profile of international terrorists with that of domestic terrorists in the United States. Smith points out that, in the past, the profile of a terrorist was based on studies of international terrorism. According to Smith, leftist groups with political agendas were the frequent perpetrators of international terrorism. However, Smith presents evidence that in the United States, terrorist activities are more commonly perpetrated by right-wing groups rather than left-wing groups. Smith reports that from 1980–1989, a total of 170 people were indicted for domestic terrorism or terrorist-related activities in the United States. Of these, 103 were associated with right-wing groups. Smith further reports that most right-wing terrorists are bound together through a shared religious ideology, Christian Identity. These groups include or have included the Aryan Nations; Sheriff's Posse Comitatus (SPC); the Order; the Order II; the Covenant, Sword, and Arm of the Lord (CSA); and the Arizona Patriots.

Smith examined the demographic characteristics of the 75 right-wing extremists indicted for terrorism, and he reported that the majority were white (97%), were male (93%), had an average age of 39 years, and tended to have minimal educational and job skills. As shown in Table 1-1, the profile and demographics of domestic terrorists today is fairly similar to that of 1994.

Table 1-1

Characteristics of Domestic Terrorists in 1994 and 2006

	Year	
Group Characteristics	**1994**	**2006**
Mean age	39	39.3
Composition by gender	93% male 7% female	87% male 13% female
Composition by race	97% white 3% Native American	100% white
Occupation	Unemployed	Manual labor
Residence	Rural	Rural
Ideology	Christian Identity	Non-Bible and pseudo-Islamic
Base of operation	Rural	Rural
Tactical approach	Camps & compounds National networking	Lone wolf Cells

As shown in Figure 1-1, more-recent data from the Rand Corporation and the Memorial Institute for the Prevention of Terrorism (MIPT) suggest that right-wing extremists continue to have the largest presence within the United States and pose the greatest threat.

Given the preponderance of right-wing domestic terrorists in the United States, this book focuses on them as a major source of terrorism in this country. The current chapter focuses on one such group, the Aryan Nations, and examines characteristics of its members.

Past research on the Aryan Nations has focused on the violent nature of the group (Flynn & Gerhardt, 1989; Hamm, 2001; Ridgeway, 1995) without providing an in-depth look at the characteristics of other members. For example, sociologist Mark Hamm (2001) studied the Aryan Republican Army, a splinter group of the Aryan Nations, and discussed

Figure 1-1

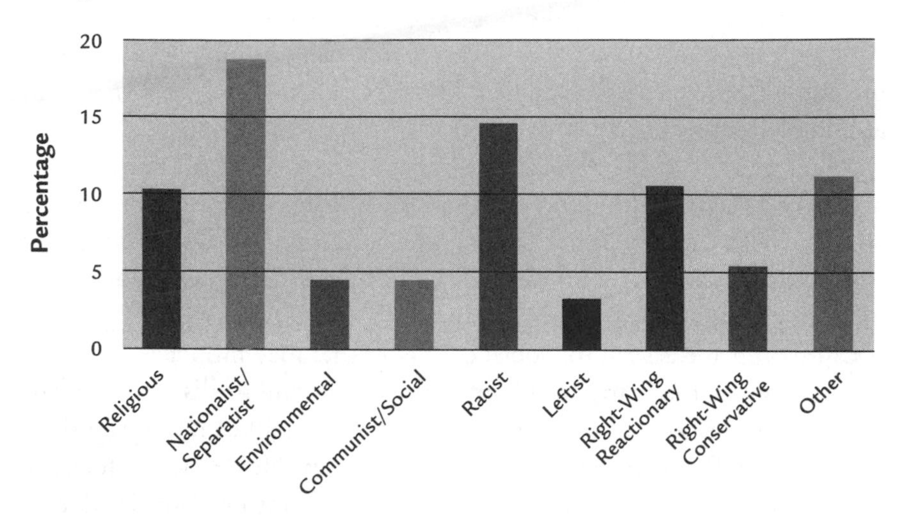

Data from: MIPT Terrorism Knowledge Base. "Incident analysis wizard." Retrieved March 3, 2006, from http://tkb.org/chwiz4.jsp. Note: Data now available through the Global Terrorism Database at http://www.start.umd.edu/data/gtd.

how this group committed bank robberies in order to finance domestic terrorist acts for future groups of the radical right-wing movement known as Christian Identity. Most of Hamm's (2001) analyses framed members as domestic terrorists, focusing on the violent nature of these individuals, without placing them into a larger framework of all Aryan Nations members.

James Ridgeway's (1995) book *Blood in the Face* examined the historical beginnings of the Aryan Nations in Idaho. However, the main focus of his analyses centers on a string of violent Brink's truck robberies, committed by a group who called themselves the Order. The Order was founded by Robert Mathews, and consisted of white supremacists who believed that the Northwest section of the United States should secede from the nation and become a territory for only whites. Although all of the individuals who committed the crimes met in Idaho at the compound of

Richard Butler, the founder of Aryan Nations and a prominent member in Christian Identity, most did not share the beliefs of Christian Identity. Although these acts of violence occurred, they were committed by splinter offshoots from the main group of Aryan Nations, and as such may not accurately reflect the beliefs or practices of the parent group.

Aho (1990), in *The Politics of Righteousness*, focused on cultural elements of Christian Identity members and discussed the demographic characteristics that were influential in bringing people to Christian Identity, the religious beliefs of the Aryan Nations. These beliefs are based on the writings and teaching of Richard Butler. Aho's (1990) book—because of its focuses on the social correlates that may influence people to become part of the Christian Identity movement and its discussion of the political motivations of those who join—makes significant contributions to understanding cultural elements of the group. However, the focus of his book is on Christian Identity, and he does not explore the demographic characteristics of the larger membership of the Aryan Nations.

Hamm (2001), Ridgeway (1995), and Aho (1990) each present a distorted view of the Aryan Nations. Hamm and Ridgeway, by focusing on the violent individuals and violent acts, paint a picture of all Aryan Nations members as domestic terrorists who support an explicitly violent ideology. Aho, although he explores cultural elements, focuses specifically on the element of Christian Identity rather than on the Aryan Nations as a whole. Because of the limited perspective, each author presents a picture of the Aryan Nations as a homogeneous group whose members have similar backgrounds and motivations.

The purpose of this chapter is not to deny the social construct of Aryan Nations members as domestic terrorists—there is, after all, an element of truth to this. Some people join the Aryan Nations or similar groups for the purpose of engaging in violence. However, there are several other reasons why people join this group. This chapter explores the variability of backgrounds and beliefs among Aryan Nations members. Specifically, through the use of interviews with members of the Aryan Nations, the demographics, ideologies, and culture of the group are examined. A better knowledge of the demographic characteristics of the group and a clearer understanding of why people join the group will provide information that may be useful in predicting not only who will join such a group, but also—of those people who do join—predicting who is most likely to commit a deviant act.

CURRENT DEMOGRAPHIC CHARACTERISTICS

Socioeconomics

As shown in Table 1-2, approximately 75% of Aryan Nations members reported that their parents were from the working class, whereas the remaining 25% reported that their parents were from either the middle or upper class.

Table 1-2

Parents' Social Class

Social Class	Percentage
Working class	78.3
Middle class	17.4
Upper class	4.3
Total	100

Given the parents' demographic information, coupled with the fact that a large percentage of people in the military come from working-class families, it is not surprising that 30% reported having served in the military, and 70% reported having no military experience. However, none of those with armed-services experience were career military. Regarding military experience, the majority of those interviewed stated that it was during their military service that they began to develop distrust for the government. This eventually led them to join the Aryan Nations. Specifically, it was in the military that most members "saw that the government was run by Jews." Many of those interviewed reported that it was because of their belief in a "Jewish conspiracy" or "Jewish control of the government" that they began their quest for the answer to the question "Why are there so many Jews in high places?" Eventually, the questioners found the answers they were seeking in the Aryan Nations.

As shown in Table 1-3, the occupational demographics of the group members are similar to those of their parents. Approximately half of those who join the Aryan Nations movement tend to have manual labor jobs.

Less than 10% reported working in a white-collar profession. Having a high percentage of members from the working class is consistent with membership in other hate movements (e.g., skinheads), and it also is consistent with Smith's (1994) findings that only 12% of right-wing terrorists had college or university degrees and a third had not completed high school.

The current data for the Aryan Nations also reveal that 26% of the group's members are unemployed, and 9% are collecting Social Security Disability Insurance (SSDI) benefits. Collecting SSDI contradicts the work ethic that is part of the movement and runs counter to the philosophy of Richard Butler, the founder of the Aryan Nations. Butler stated that—because the Aryan Nations does not support the way the government is run—the group's members should not be reliant on the government for assistance (Alibrandi & Wassmuth, 1999; Ezekiel, 1996). Given that many Aryan Nations members believe that "the Jews run the government," then, by taking money from the government, these members are essentially being supported by the people they vilify.

Table 1-3

Occupation

Occupation	Percentage
Manual labor	52
White-collar profession	9
Own business	4
Unemployed	26
Collecting SSDI	9
Total	100

Age

Previous researchers have suggested that hate-group membership appeals predominately to young males (cf. Blee, 1996; Blee, 2002; Hamm, 1995; Smith, 1994). Whereas past research has been correct in surmising that youths are attracted to the violence, symbols, and hate rock of such groups as skinheads, these findings do not apply to Aryan Nations members.

Ideology and recruitment for the Aryan Nations is not based on hate rock—which, as Mark Hamm (1995) has shown, is the strongest predictor for youths joining this subgroup. Instead, the Aryan Nations places emphasis on justifying their hatred through religion, specifically the biblical teachings of a radical religion called Christian Identity. Christian Identity teaches that the Jews are related to the devil, and that their sole purpose for being placed on the earth is to destroy Judeo-Christianity and to take over the world by controlling the major social institutions in the United States (see Barkum, 1996).

Because religious ideology is the basis for the Aryan Nations and is practiced less by youths than by adults, one would predict that the Aryan Nations would have a wider appeal to an older adult population. Consistent with this prediction, the average age of the demographic under study was 39.3 years, suggesting a more middle-aged population. These findings also are consistent with those reported by Smith (1994), that the average age of right-wing American terrorists was 39.

Gender

Past research suggests that hate-group members are predominantly male (see Blee, 2002; Hamm, 1995; Smith 1994). Consistent with this research, 87% of the Aryan Nations membership was male, and 13% was female. This figure is higher than that reported by Smith. However, the difference may result from the fact that Smith's sample was limited to individuals who had been indicted. The small percentage of females may be due to the patriarchal nature of the group. Women play a small role within the movement and are sometimes excluded from attending meetings. Aryan Nations members believe that women should dress in traditional, conservative fashions. Women who have an active role within the organization tend to fulfill stereotypical, supportive functions such as cooking or running the women's auxiliary league—which, in this case, is purely symbolic because the league has no members.

Most members of the Aryan Nations, both male and female, support a patriarchy and believe that a woman's role is to stay home, care for the children, and tend to her "husband." One participant stated: "I don't believe that women should be in the work force. The Bible states that women are the caretakers of the family. I think that women should concentrate on having as many babies as possible so the white race does not become extinct."

Marriage and Family

Because Aryan Nations members believe in returning to family values, one would expect a large percentage of the group's members to be married and have large families. As Table 1-4 shows, however, less than half of those interviewed were either currently married or involved in a common-law marriage. Almost a third of the group had never been married.

Table 1-4

Marital Status

Marital Status	Percentage
Never married	30
Married	30
Divorced	17
Common Law Marriage	14
Widowed	9
Total	100

When asked why so many in the group were not married, one member explained:

> Society has stigmatized us. Most of the general public thinks that we are a bunch of weirdoes [*sic*] who want to kill everyone. I think that most women believe what they hear on television and do not want to get mixed up with anyone who is in the movement, for fear of their lives.

The Web site http://www.Stormfront.org/forum, a "White Nationalist Community," as part of its forum, has a section *White Singles* that features *Talk and Dating Advice*. Responses to the question "Does being a White Nationalist has [*sic*] an impact on dating?" suggest that some group members feel stigmatized. For example, one respondent wrote:

> Wearing your own personal political identity on your sleeve generally signals most females that your [*sic*] an extremist or fanatic and "may" cause them to think you are misguided, ill informed or worse—insincere and looking for attention.

However, others suggested it was important to be up-front about your beliefs: "First and foremost, never suppress your beliefs." And further, that being a White Nationalist required greater selectivity in dating partners: "Yes . . . NO non whites allowed. No race traitors. No race mixers. European descent is the only thing authorized."

Similarly, responses to the question "How do you address your politics in [the] first couple of dates?" suggest the importance of being up-front about your beliefs. One respondent wrote:

> Basically its [*sic*] good to get your feelings out within the first month of a new relationship. Its [*sic*] perfectly fine to ease someone into your views instead of hitting them over the head with the whole bag at once. But dont [*sic*] compromise your beliefs for anyone.

Another person wrote: "I've been as blunt as saying 'I hate niggers, how about you?'"

One reason for the high percentage of common-law marriages is that some members do not believe in the governance of society and would not get married by a "Judeo-Christian church." Although some couples in common-law marriages could be married by a Christian Identity preacher, many chose not to because they did not want to lose child support from previous relationships.

Consistent with the focus on family as well as the patriarchal beliefs of the organization, Aryan Nations members believe that the most important function of a woman is childbearing. By emphasizing childbearing, members believe that they are adding to the population of the white race in order to offset the browning of America. One participant stated:

> The white race is going to become extinct soon. Soon you will see muds (Blacks) taking over the world and we will have no rights left. I think every white person has a right to bring as many white children into this society so we can prevent that from happening.

One female forum member (http://www.Stormfront.org/forum) describes herself as "Breeding for the continuation of *my culture*, survival of *my people*, and love of *my unique race*. Having WHITE children is a form of warfare that WHITE women are built for."

Consistent with this belief, postings on www.Stormfront.org/forum under *Homemaking* suggest there is a great deal of support for whites having children. Members announce when they are expecting or recently had a baby, and congratulations are offered. In response to the inquiry "Anyone here encouraging more babies from fellow whites?" one respondent wrote:

> I feel as white nationalists are [*sic*] duty should be to do this more often, encouraging at least 1 more white baby to be born to try and reverse this god awful trend where white people lose their evolutionary focus of staying alive in the future while mestizos continually brag about having tons of kids and not teaching them english [*sic*].

Child rearing is so important to the group that women are given a medal if they have more than five children. However, contrary to this emphasis on family, very few of those interviewed had large families (see Table 1-5). The data reveal that only 22% of the members involved in the organization had children. Of those with children, most had relatively small families. Only 1% reported having five or more children.

Table 1-5

Percentage of Members by Number of Children

Number of Children	Percentage
0	78
1	0
2	17
3	4
4	0
5 or more	1
Total	100

In addition to having children, the discussions on Stormfront.org/forum indicate that child rearing—specifically, raising children to share your beliefs—is important. In response to one individual who was expecting another baby, one person said, "Raise him/her the white way." In addition to the discussion site for *Homemaking*—which includes gardening, cooking, and child rearing—there also is a specific discussion forum for *Education and Home Schooling*.

Social Influences

When participants were asked why they joined the Aryan Nations, 21 of the 24 (87.5%) interviewees reported that they had been involved with other hate groups prior to their membership in the Aryan Nations. On average, participants reported being in 2.3 hate groups prior to joining the Aryan Nations. This finding suggests that these members are "lifers" in the hate movement, perhaps searching for the hate group whose values are most consistent with their own beliefs. This finding is contrary to previous research (Ezekiel, 1996), which suggests that membership in hate groups is transient and reflective of people "dabbling" in hate as they explore the "evils" of blacks and Jews. As shown in Table 1-6, when asked why they had joined the Aryan Nations, 39% reported that their friends had introduced them to the organization and had influenced their joining.

Table 1-6

Source of Influence for Joining Aryan Nations

Source of Influence	Percentage
Friends	39
Environment	30
Violent image	18
Prison	9
Internet	4
Total	100

Through the course of the interviews, it became obvious that most members feel strongly about their beliefs and do not see what they are doing as wrong. These individuals believe they are the victims of society and have a right to preach their hatred. In fact, almost one-third of those interviewed reported that they joined the Aryan Nations because of the environment or current state of society. A frequently heard response was "I didn't want to become a racist; society made me that way." Most members believe that the preferential treatment being given to blacks, Jews, women, and other minorities is destroying society—and that they need to do something about it. In the end, their response was to join the Aryan Nations.

A smaller proportion of those interviewed reported that it was while they were in prison that their "eyes were opened" and they joined the Aryan Nations. These respondents reported that in prison, they learned how to hate, and further, "that the Jew was the sole [*sic*] of society's problems."

Somewhat surprisingly, exposure to hate on the Internet was the reason least frequently mentioned for joining the Aryan Nations.

Other members reported that they were attracted to the Aryan Nations because of the group's violent image and its willingness to use violence. These members felt that this was the "advocacy" group they were looking for, for white rights. These individuals have a propensity for violence. They felt that the Aryan Nations, unlike the National Alliance or the Ku Klux Klan (KKK), was actually working to spread their hate—and if the group's work resulted in violence, so be it. To these individuals, violence is seen as a means to an end, something that will help fix the problems of society.

Whether individuals were recruited into the Aryan Nations by friends or through prison networks, a recurrent theme throughout is the belief that the source of their difficulties (e.g., unemployment, poverty, etc.) stems from the actions of the government, through its preferential treatment of minorities and women. Thus, many of these individuals feel "forced" into becoming racist in order to protect themselves and their families and they feel justified in using violence to this end. These beliefs are consistent with the findings that perpetrators of hate often see themselves as the victims, and consequently feel justified in their hatred and violence toward those who *persecute them* (Baumeister, 1996; Post, 1999).

Criminal Involvement

As mentioned previously, some participants reported that the violent nature of the Aryan Nations appealed to them. When asked about criminal behavior or involvement in a crime, slightly less than half (43%) of those interviewed reported having been arrested for a crime, whereas 57% reported that they had not been involved in a crime. Of those who had engaged in a crime, the two most frequent reasons provided for their criminal activity were either to get back at what they feel is a Jewish-run society or to help fund the movement. The types of crimes engaged in vary from misdemeanors to felonies. Of those respondents who were arrested, crimes included attempted murder, murder, assault, drinking and driving, weapons violations, check forgery, and bank robbery. Several people were arrested for multiple drug offenses, usually involving drug sales.

Regarding gun ownership, only 26% of the Aryan Nations members who were interviewed reported owning a gun, and 74% reported that they did not own one. Although this figure may seem low, it may be due in part to the fact that a high percentage of members (43%) have committed major crimes and are mandated by state and federal laws to be ineligible for gun ownership.

Geography

Whereas past research on hate groups such as the Aryan Nations, KKK, and skinheads has focused on the psychology or sociology of these groups, little research has examined the geographic location of these groups or a group's reasons for locating in a specific area. In the past, hate groups were concentrated in the South, and, for this reason, these groups tended to be viewed as a Southern issue. However, during the past several decades, hate groups have spread across the United States and are now present in 46 out of 50 states (Intelligence Project, 2004). Although most hate groups are located in urban or suburban areas, their presence is increasing in rural areas. As shown in Table 1-7, more than half of those interviewed lived in rural areas.

Table 1-7

Geographic Location

Location	Percentage
Rural	54
Suburban	33
Urban	13
Total	100

One important goal of hate groups located in urban and suburban areas is to confront the enemy. Typically, hate groups located in cities with a high percentage of minorities work to either drive the minority groups out of the area or to convince other whites that minorities are physically, mentally, and biologically different from whites. The hate groups use various forms of technology, such as public-access television or shortwave radio broadcasts, to get their message out. Frequently, members of hate groups will visit college campuses to leave hate material in libraries and on cars.

In the past decade, several chapters of the Aryan Nations have made a conscious decision to move their operations from urban to rural areas.

These moves often stem from the realization that they could not drive the minority groups from the cities. These chapters would then select predominantly white communities, typically with stagnant agrarian economies, in which to settle and preserve their white community.

The rural groups also use technology to let like-minded others know that they have settled in a community. Frequently, the media—by providing extensive coverage of a hate group's movement into a rural area—broadcasts the group's new location to an audience larger than would have been achieved by the hate group alone. Consequently, the media may unintentionally facilitate the hate group's goals of not only letting others—both hate-group members and minority-group members—know where they have settled, but also facilitating the growth of an all-white community by advertising who is welcome (and, conversely, who is unwelcome) in the community.

CONCLUSIONS

In summary, the results of these interviews suggest that members of the Aryan Nations tend to be males in their late 30s. As stated previously, it is typical of hate groups to have a predominantly male membership. The average age of Aryan Nations members is higher than that in other hate groups. One possible explanation for this is that the religious focus of the Aryan Nations might be less appealing to adolescents and young adults than are the hate music or violence associated with other hate groups. The majority of Aryan Nations members reported coming from a working-class background and, as a source of employment, engaging in manual labor. Given the group's strong work ethic, it is somewhat surprising that approximately 25% of those interviewed were unemployed, and that almost 10% were collecting SSDI benefits.

The Aryan Nations advocates marriage and family, promoting the belief that it is the duty of white people to have children to offset or forestall the "browning of America." Slightly less than half of those interviewed were currently married or in a common-law marriage, whereas 30% reported having never been married. One individual—perhaps in an attempt to negate personal responsibility for not having a partner—blamed society for his lack of success in finding a spouse. He stated that because the Aryan Nations has such a bad reputation, it was difficult to find a partner. Also, contrary to the emphasis of the Aryan Nations on family, less than 25% of those interviewed had children, even though most believe that "the future of the white race belongs to the children."

As stated in the introduction, the stereotypical image of an Aryan Nations member is that of a violent criminal. Slightly less than half of those interviewed had participated in a crime. Future research needs to examine what motivates some members to engage in criminal activity. Interestingly, only a fourth of those interviewed reported currently owning a gun. However, one possible explanation for the low percentage of gun ownership is that, for those who had committed a crime, state or federal laws may have made them ineligible to own a gun.

Finally, none of the people interviewed were reared to be racists. Consistent with research by Kathleen Blee (1996, 2002), the current research demonstrates that individuals are introduced to the hate movement by others, learning to hate minorities and Jews through their social networks. Of those interviewed, 40% suggested they were recruited into the Aryan Nations through friends. Approximately one-third joined the group as a way of combating societal ills, specifically the preferential treatment given to minorities and women. Interestingly, almost 90% of the people interviewed reported previous involvement in other hate groups. This finding suggests that these individuals are not merely dabbling in hate, but are lifetime members who are shopping for the organization that best matches their beliefs. Although these individuals may become disenchanted by the message of one hate group, rather than abandoning such groups altogether, they look for another hate group that can justify their hate with a different and acceptable message.

REFERENCES

Aho, J. (1990). *The politics of righteousness*. Seattle: University of Washington Press.

Alibrandi, T. & Wassmuth, W. (1999). *Hate is my neighbor*. Ellensburg, WA: Stand Together Publishers.

Barkum, M. (1996). *Religion and the radical right*. Chapel Hill: University of North Carolina Press.

Baumeister, R. F. (1996). *Evil: Inside human violence and cruelty*. New York: Freeman.

Blee, K. (1996). "Becoming a racist: Women in contemporary Ku Klux Klan and neo-Nazi groups." *Gender and Society* 10: 680–702.

Blee, K. (2002). *Inside organized racism: Women in the hate movement.* Berkeley: University of California Press.

Ezekiel, R. (1996). *The racist mind: Portraits of American neo-Nazis and Klansmen.* New York: Viking.

Flynn, K. & Gerhardt, G. (1989). *The silent brotherhood: Inside America's racist underground.* New York: Free Press.

Hamm, M. S. (1995). *American skinheads: The criminology and control of hate crime.* New York: Praeger.

Hamm, M. S. (2001). *In bad company.* Boston: Northeastern University Press.

Intelligence Project. (2004). "Active hate groups." Retrieved June 6, 2005, from http://www.tolerance.org/maps/hate.

MIPT Terrorism Knowledge Base. "Incident analysis wizard." Retrieved March 3, 2006, from http://tkb.org/chwiz4.jsp.

Post, J. M. (1999). "The psychopolitics of hatred: Commentary on Ervin Staub's article 'Peace and conflict.'" *Journal of Peace Psychology* 5: 337–344.

Ridgeway, J. (1995). *Blood in the face.* New York: Thunder Mountain Press.

Smith, Brent L. (1994). *Terrorism in America: Pipe bombs and pipe dreams.* Albany: State University of New York Press.

Stormfront.org. "Forum." Messages retrieved April 12, 2007, from http://www.stormfront.org/forum.

CHAPTER 2

The Criminology of Terrorism: Theories and Models

Thomas R. O'Connor

INTRODUCTION TO THEORY

This chapter discusses the major criminological theories of terrorism, or those theories that offer insights into what might be called explanations of terrorist behavior as criminal conduct. Explanations, as opposed to understandings, are designed to provide scientifically valid guides to research, and sometimes policy. Theory-driven policy is rare, but far more common is theory-driven research. A good theory has practical uses if it provides, for example, some fairly clear ideas about what are the independent variables (the causes) and the dependent variables (the effects). Criminological theory is mostly about attempting to correctly identify the causes of behavior, and this quest often takes us into the realm of concepts we cannot see or observe directly. It helps if the concepts in a theory have clear meanings so researchers can test them by collecting and analyzing data that prove or disprove a theory. Admittedly, not all theories are inherently testable, and few offer crystal-clear meanings, but the job of explanation is never easy. By reviewing those theories that have stood the test of time, this chapter hopes to encourage more interest in theories of terrorism, the development of which is urgently needed.

Most of the major theories of terrorism are derived from theories of collective violence in the field of political science. Indeed, prior to the emergence of criminal justice as a separate discipline in the early 1970s, it can be safely said that political science pretty much had a monopoly over theories of terrorism, followed perhaps by the disciplines of religion and economics. Sociological, psychological, and criminological theories also have certainly had a role to play with some relevance. We will begin, first, with the theories of political violence. It is customary to say at this point that none of the following ideologies—or any ideology, for that matter—are being advocated. The purpose is to provide an objective overview of theoretical concepts, causal factors, and models connecting cause and effect. The underlying concern should be to answer the questions "Why does terrorism occur?" or "What causes it?" rather than pass judgment or assess any of the theories at this point. With the political theories, it often is the case that the form of governance is held to be the main cause of terrorism, and with the other theories, a number of subcultural and personality factors are usually found to be at work. With other theories, such as sociology, the interplay between social movements and societal response is often looked at to help explain terrorism.

It should be noted that theory is more than the study of motive. In criminology, theories tend to take on more than the explanation of offender mind-sets and behavior. They often tackle issues such as victimology and criminal justice system response. Motive itself is frequently ignored in the prosecution of terrorists (Smith, 1994; Smith, Damphousse, Jackson, & Karlson, 2002). Justice itself often is evasive. Sentencing variations in bringing terrorists to justice occur mostly because of "structural-contextual" effects (when components of criminal justice work at odds with one another), and also because of "liberation" effects (when judges and juries nullify the law or follow their own sentiments). Inconsistencies in sentencing affect the ability to collect research data because an unindicted or acquited terrorist is not, legally, a terrorist. For this reason, terrorism databases, such as the one at the Memorial Institute for the Prevention of Terrorism (MIPT) Terrorism Knowledge Base (www.mipt.org), are often incomplete. However, analyses using such databases afford opportunities to test theories, tease out relevant variables or factors, and do comparative theory testing, and so on—all of which point to the larger purposes of theory. It should be remembered, as the following theories are reviewed, that a strict legalistic conception of who is or is not a terrorist makes for a more rigorous approach. Just because somebody shares some of the ideas mentioned as causal factors below does not make them a terrorist. A court of law determines if a person is a true terrorist, and any theory that can explain court behavior as well as terrorist behavior definitely has some added value.

THE POLITICAL THEORY OF ANARCHISM AS A THEORY OF TERRORISM

Terrorism is most definitely not a form of governance, but anarchism is. Most anarchists reject terrorism in its vanguard varieties (for nationalist or religious purposes), but in a theoretical sense, anarchism justifies terrorism as a form of criminal action that attacks the values of an organized, complacent society. Anarchism is a theory of governance that rejects any form of central or external authority, preferring instead to replace it with alternative forms of organization such as shaming rituals for deviants, mutual assistance pacts between citizens, syndicalism (any nonauthoritarian organizational structure that gives the greatest freedom to workers), iconoclasm (the destruction of cherished beliefs), libertarianism (a belief in absolute liberty), and plain old rugged individualism. Anarchism often is referred to as the 19th-century roots of terrorism, the term first being introduced in 1840 by Pierre-Joseph Proudhon. Anarchism is defined as the rejection of the state, of any form of coercive government, and of any form of domination and exploitation. It involves the notion of free and equal access to all the world's resources to enable positive freedom (freedom to) in place of negative freedom (freedom from, or the basis of most constitutional rights).

As a theory, anarchism holds a unique place in history because it was the first revolutionary movement to come up with systematic ideas about the purpose of agitation. The reader will surely recognize some of these ideas as terrorist tactics, but it is important first to understand them in the context of anarchism. Proudhon contributed the idea of finding the "moment," as when the moment is ripe for revolutionary action. Another anarchist, Mikhail Bakunin, popularized the idea of "propaganda by deed," or letting your actions speak for themselves—which was a theory originally developed by Carlo Pisacane, an Italian revolutionary who argued that ideas spring from deeds and not the other way around. Over the years, this notion has evolved into a fairly coherent philosophy of the bomb as part of a propaganda campaign to stimulate awareness and sympathy with the cause, and in this respect has been noted as a defining feature of terrorism (Georges-Abeyie & Hass, 1982). Bakunin's ideas strongly influenced anarchism because his concept of propaganda by deed also included a prohibition against large-scale group action (it being better, he thought, for anarchist action to be individualized or done in small groups). Most anarchists operate on the principle of leaderless resistance, or acting on your own, with little knowledge or support of the groups they may belong to. Another anarchist, Sergei Nechaev, who was an associate of Bakunin's, glorified the "merciless" aspect of destruction, but it was Bakunin who laid out the six steps necessary to destroy a social structure, as paraphrased below:

- Kill the intelligensia (kill those who are intelligent and most powerful in society)
- Kidnap the rich and powerful (those who yield the biggest ransoms)
- Infiltrate the politicians (to find out their secrets and discredit them)
- Help the guilty criminals (to confuse society over justice and punishment)
- Defend the loudmouths (those who make dangerous declarations)
- Nurture the supporters (help fellow travelers who believe in societal destruction)

Major anarchist figures, such as Karl Heinzen and Johann Most, contributed the idea that murder, especially murder-suicide, constituted the highest form of revolutionary struggle. Both advocated the use of weapons of mass destruction (WMD). Other anarchists contributed additional ideas, such as Pyotr Kropotkin's notion of "propaganda by word," or radicalizing the public by use of subversive publications. Anarchism (like fascism) also has had some influential female figures, such as Emma Goldman (1869–1940), who comes to mind as an early founder of free speech (the American Civil Liberties Union, or ACLU) and sexual freedom movements. Minor figures in the history of anarchism—such as Charles Gallo, Auguste Vaillante, Émile Henry, and Ravachol (François Claudius Koenigstein)—advocated the idea that to have the most effect, the targets must be innocents (for example, in places such as crowded dance halls or shopping centers) or symbols of economic success (for example, banks or stock exchanges). It may be worth noting, in passing, that the famous Italian criminologist Cesare Lombroso developed his notion of the "born criminal" in part by being called in to examine the physical features of some minor anarchists who were really nothing more than criminals justifying their behavior with anarchist talk.

Between 1875 and 1912, anarchists alone or in small groups managed to assassinate or attempt to assassinate the leaders of nine (9) different countries, including the United States (President William McKinley in 1901). These crimes were just the best-known acts of anarchism because anarchists also were involved in numerous ordinary crimes such as theft, robbery, murder, kidnapping, assault, and bombing. The most famous incident was the Haymarket riot in Chicago during 1886. During these peak years for classic anarchism, May Day celebrations became famous as all-out crime-rampant days. Police departments around the world became convinced there was an international conspiracy, and suspicious foreigners were locked up by the hundreds in many countries. Perfunctory trials were held, and many defendants were hanged or deported. The most famous of these trials was the 1920 case of Nicola Sacco and Bartolomeo Vanzetti, who were more antiwar and labor activists than anarchists. Anarchism—in the classical sense—was largely erased from the face of the earth by 1917 via a number of factors: the rise of communism and fascism (both of which are opposed to anarchism), strong xenophobic deportation (Red Scare) laws in democratic countries, and the fact that classic anarchism never became an organized movement. Twentieth-century terrorist groups that emerged later and claimed an ancestry with anarchism include the Japanese Red Army, the British Angry Brigade, the German Baader-Meinhof Gang, the Weathermen in the United States, and the Mexican Zapatista movement (Kushner, 2003). During the Spanish Civil War in the 1930s, something called anarcho-syndicalism developed, which is a loose confederation of various protest groups. Those who call themselves anarchists today (see Purkis & Bowen, 1997) are more likely to be environmentalists or part of the antiglobalization movement, and they target such institutions as the World Bank, the International Monetary Fund (IMF), or the World Trade Organization (WTO).

For purpose of balance, it is important to point out that much anarchist thought does not support terrorism. It has historically supported terrorism, and even today might support assassination, but there are only weak analytical links between the two, most strongly with the concept of propaganda by deed. Anarchists hold to a doctrine that anarchy must be created in the act of self-liberation from oppressive and coercive relationships. You don't blow up the relationship as terrorists do; instead, you convince others that grounds for the existing relationship must be blown up. Anarchism is not really about mad bombing or chaos, as it often is portrayed. Terrorists target people; anarchists target things, such as institutions and structures. Bakunin did not want the death of people, but the destruction of things and positions of authority. Only a small minority of terrorists have ever been anarchists, and only a small minority of anarchists have ever been terrorists.

In fact, there is an area of study called anarchist criminology, a controversial subfield of critical criminology that celebrates the fact that anarchism really has no workable definition (Tifft, 1979; Ferrell, 1997). Anarchist criminology advocates the abolishment of criminal justice systems. It argues that much harm has been committed in the name of reasonableness, and anarchist criminology is committed to promoting the unthinkable and unreasonable. Like other subfields of critical criminology, anarchist criminology views the state as an inherently oppressive entity, and anarchist justice advocates not only social justice (equal access to all resources), but the protection of diversity and differences among people (Ferrell, 1999).

THE POLITICAL THEORY OF FASCISM AS A THEORY OF TERRORISM

Fascism is the one form of government with the most disagreement about a definition for it. Passmore (2002) attempts a definition based on two ideas: extreme nationalism and shameless racism. The word "fascism" comes from the Latin "fasces", which means to use power to scare or impress people. It generally refers to the consolidation of all economic and political power into some form of superpatriotism toward a cult of personality figure and a devotion to endless war with one's enemies. Benito Mussolini, who practically invented the term in 1922, said that fascism is the merger of state and corporate power, but Mussolini's version was based on the idea of an indomitable power and an attempt to resurrect imperial Rome. Adolf Hitler said fascism is the clever and constant application of propaganda so that people can be made to see paradise as hell, and the other way around. Hitler's brand of fascism drew upon philosophical reflections by Hegel, Nietzsche, and Spengler and also drew upon Nordic folk romance from Wagner to Tacitus. Japanese fascism involved racism, fanaticism, historical destiny, and a mixture of Bushido, Zen and Shinto Buddhism, emperor worship, and past samurai legends.

So-called Islamofascism can be traced to the time period of the birth of Nazi "national socialist" fascism in 1928, when the Muslim Brotherhood (Al Ikhwan Al Muslimun)—parent organization of numerous terrorist groups—was formed in reaction to the 1924 abolition of the caliphate by the Turks. Islamofascism draws heavily upon Muslim Brotherhood pamphleteers, but also upon the Koran, the career of Saladin, and the tracts of Nasserites and Baathists. The term "Islamic Fascism" is a better term to use, best describing the agenda of contemporary radical extremists who happen to believe in Islam. Fascism is almost always reactionary. In one sense, it is born out of insecurity and a sense of failure, but, in another sense, fascism thrives in a once-proud, humbled but ascendant, people. Envy and false grievance are the trademarks of many reactionary movements. Believers are subject to all kinds of conspiratorial delusions that setbacks were caused by others and can be erased through ever more zealotry. Fascist leaders love conspiracies and lies—and little they say can or should be trusted.

Fascism frequently supports terrorism at home and abroad. Charismatic leaders usually are given supreme powers to crack down on dissidents, peacemakers, and anyone who does not abide by the "cult of the individual," which worships a he-man mentality and the party line. With the frequent wars and militaristic ventures that come with fascism, an effort is made to demonize the enemy as subhumans who deserve extinction. These enemies also are made into scapegoats for all a nation's past problems. Fascism appeals to the frustrations and resentments of a race of people who think they ought to have a bigger place at the global table. When combined with an anti-Western slant (e.g., the United States as Great Satan), fascism becomes a means of social identity (for example, Pan-Africanism, Pan-Arabism, and Islamofascism) as well as a facilitator of antiglobalistic terrorism.

Frustrated fascists who fail to gain control in their own countries have historically turned to terrorism. They are quite likely to turn to state terrorism or genocide because fascists do not believe that citizen rights are bestowed merely because someone inhabits a country. In other words, they do not believe that all human beings are possessed of equal rights—not only within their own country, but in the rest of the world as well. "Foreign" families and businesses (as fascists define them) usually are targeted for extermination by fascists. The enemies who are seen as the greatest threat often are those whom fascists see as corrupting or poisoning family and property relations.

Fascism is full of ironies and contradictions. On the one hand, it is antimodern in its glorification of the land, return to country life, and fascination with peasant dress or costume. On the other hand, it is pro-modern in its worship of military technology, favoritism of big business, mass mobilization of people, promotion of commercialized sport, and surprisingly liberal attitude toward the involvement of women in the movement. Science and scholarship also take on interesting twists under fascism. "Hard" sciences such as biology and chemistry usually advance significantly, especially in areas such as genetic research. "Soft" sciences such as sociology and psychology usually become usurped into mumbo-jumbo pseudoscientific talk about a glorified folk culture and reasons for hating the enemy.

Just as anarchists have their May Day (May 1) celebrations, fascists also tend to celebrate anniversaries. Many terrorists, of course, have been known to time their attacks to coincide with some date for a historical event or the birthday of someone special to them. For example, with ecoterrorism, that day is October 16—which coincides with United Nations World Food Day—and is usually when fast food restaurants are targeted for vandalism. However, the most important date in the history of terrorism is April 19. A number of significant events have happened on that date. Right-wing domestic terrorist groups call it "Militia Day" because it was when the siege at Waco ended, it was when surveillance began at the Ruby Ridge compound in Idaho, and it marks the anniversary of the Oklahoma City bombing of the Murrah Federal Building. Neo-Nazi fascist groups celebrate April 19 because it was the day German Nazis started wiping out Jewish ghettos across Europe, as well as the following day being Adolf Hitler's birthday. Internationally, terrorist groups who regard themselves as "freedom fighters," and trace at least part of this justification to the American Revolution, take heart in the fact that the American Revolution started on April 19, 1775, at the Battle of Lexington.

THE PHILOSOPHICAL THEORY OF RELIGION AS A THEORY OF TERRORISM

More than one criminologist has pointed out that the disciplines of theology, religion, and philosophy have had important things to say about terrorism (Stitt, 2003; Kraemer, 2004). It also is a fact that about a quarter of all terrorist groups and about half of the most dangerous ones on earth are motivated primarily by religious concerns (Hoffman, 1993). They believe that God not only approves of their actions, but that God demands their actions. Their cause is sacred and consists of a combined sense of hope for the future and vengeance for the past. Of these two components, the backward-looking desire for vengeance may be the more important trigger for terrorism because the forward-looking component (called apocalyptic thinking, or eschatology) produces wild-eyed fanatics who are more a danger to themselves and their own people. The trick to successful use of terrorism in the name of religion rests upon convincing believers or convertees that a "neglected duty" exists in the fundamental, mainstream part of the religion. Religious terrorism is therefore not about extremism, fanaticism, sects, or cults, but is instead all about a fundamentalist or militant interpretation of the basic tenets. Most religious traditions are filled with plenty of violent images at their core, and destruction or self-destruction is a central part of the logic behind religion-based terrorism (Juergensmeyer, 2001). Evil often is defined as malignant narcissism from a theological point of view, and religion easily serves as moral cover for self-centered terrorists and psychopaths (Stitt, 2003). Religion has always served the function of absorbing or absolving evil, or at least assuaging guilt in what theologians call theodicy, or the study of how the existence of evil can be reconciled with a good and benevolent God. Most religions theodicize away evil as either: (1) a test of faith; (2) a product of free will; (3) part of God's plan; or (4) a lesson to let people learn right from wrong. The use and misuse of these theodicies are the most common pathways that religious terrorists follow in the early stages of their behavior (Kraemer, 2004).

To be sure, another usual pattern in religious-based terrorism is for a psychopathic, spiritual leader to arise who is regarded as somewhat eccentric at first (a tendency toward messianism). But then, as this leader develops their charisma, they tend to appear more and more mainstream and scholarly. They begin to mingle political with religious issues (a tendency toward theocracy), and little-known religious symbols or pieces of sacred text take on new significance. Quite often, these symbols are claimed to be an important part of that religion's history that has somehow been neglected. The stage is then set for blaming somebody for the betrayal of this sacred heritage. First, the politicians in one's own country are blamed, but soon a foreign influence, such as secularization or modernization, is blamed. Militant religions quickly move to blaming a foreign influence for at least three reasons: (1) it does not serve the religion's survival interests to blame a homeland; (2) blaming makes use of a long history of competition, animosity, and war between the world's different religions; and (3) any blaming to be done must occur on the symbolic or cosmic level, which is to say that the enemy cannot have a face, but must be some impersonal, evil-like force or influence. Hence, the most specific enemy a militant religion can have is some global trend such as secularization, modernization, or Westernization. The strength of messianic power is its ability to guarantee that a radical change is coming without specifying exactly what that change will look like. However, once a semivague enemy has been identified, the religious movement borrows the idea of "sovereignty" from the political realm, and begins to see itself as the legitimate defender of the faith and the only restorer of dignity to the homeland. Most importantly, such "defenders" justify terrorist action by their accountability only to God, for it is God who has chosen them for this sacred mission in history.

Perhaps the most interesting aspect of religion as a theory of terrorism is how a devout believer could come to mix politics and religion in the ways just described. The answer is in the conception of worship. Most people associate worship with dressing up, the ringing of church bells, and explicit one-way communication with God (human to God). But worship is not just about the liturgy of church ritual. Worship is part of service to God and all of humanity on behalf of God (striving to receive instructions from God). Politics also is about service, especially public service. So-called "liberation theology," which permeates much of Latin America, has always had a handle on this aspect of worship as service that is intended to bring about the emancipation of the poor. Antonio Gramsci (1891–1937)—a founder of the Italian Communist Party who is best remembered for the concept of hegemony (Bocock, 1986), or the idea of an all-encompassing worldview—once postulated a model of worship as opposition. To engage in any sort of enterprise involving service to God, humanity, or social justice, each group of devout worshippers must see through their religious culture toward political goals. It is not so much using religion to achieve secular ends, but rather, the transformation of theology to create "free spaces" that permit creative action consistent with that religion's view on the needed transformation of society. A key theological transformation that supports terrorism, then, is the notion that communal violence, even though violence is despised, is a form of worship that may help discover the true nature of God and open up two-way communication with God (human to God and God to human).

Religious terrorism can be quite extreme in its tactics. Not only does it strive to avenge a long history of persecution and injustice, but it frequently carries out preemptive attacks. This is because a high level of paranoia is usually maintained about the actual degree of threat that the enemy trend poses. Rarely are religious terrorists swayed by secular sources of information about the degree of actual threat, but instead are driven by doctrinal differences of opinion over interpretations of scriptures and symbols. This results in two things: (1) a rather nonselective targeting pattern, lashing out blindly, often harming innocents; and (2) the creation of numerous offshoot, spin-off, or fringe groups who believe they are commanded to follow a somewhat different mission imperative. Add to this the fact that most adherents have already long felt like alienated and marginal members of society, and you have a recipe for perhaps the most dangerous and prolific kind of terrorism in the world today.

Most religious terrorist groups can trace their origins to key historical events. Institutional memory is long, as the example of Irish terrorism points out, and it is not uncommon for the group to create rituals designed to "never forget" some long-ago grievance. In one sense, this is why religious terrorism is popular—because political terrorism, like politics, has a much shorter memory. Another variety of religious terrorism has its roots in millenarianism, where the key event is some doomsday or apocalyptic date when something important is supposed to happen. It is known from studies of UFO cults that such groups often are more dangerous after an event fails to happen. This is because of cognitive dissonance, which forces a rearrangement of attitudes and beliefs that are frequently more rigid and fantastic than before. However, political events also serve as the catalyst for religious terrorism, and these are usually tied into whatever messianic traditions the religion has. For example, the rise of Al Ikhwan Muslim militancy can be traced to a date in 1979 (during the Islamic year 1400), when the return of the prophet Mahdi was anticipated at the Grand Mosque in Mecca. Adherents of this belief stormed the mosque by force that year, which happened to coincide with a time of pilgrimage and the height of the tourist season. The government reacted by forcing the militants out, cementing forever a date of infamy in which the group became certain that the homeland needed rescuing from secularization. Religious terrorists also typically have "mourning periods" or dates such as "anniversary of the martyrs" because these activities are important ways the group recruits new members who have been standing on the sidelines. Recruitment generally is followed by a reeducation program that changes the way a person thinks about good and evil. Anything foreign, secular, or modern without question becomes evil; anything supporting an all-out, uncompromising struggle against the enemy, including the killing of innocents, becomes good. The only exceptions are when a group has freed up some nonviolent avenues of experimentation.

It is important to understand the practice of martyrdom in the terrorist context. Not only does a martyr serve recruitment and other purposes after their death, but a whole mythology develops around them, which might be called a process of martyrology (Ranstorp, 1996). Targets are chosen not for strategic purposes but for symbolic purposes, and the repercussions of an attack are managed as well. The ideal target is one in which the martyr can inflict more damage than is expected for their size. The idea is to produce an impression that the group is larger and more powerful than it actually is. This feeling of power is enhanced by the use of anonymity, whereby the martyr goes through an indoctrination process where they are stripped of their real identity and provided with a false background history. The process goes much further than establishing a cover story in case of capture. The process involves changing the family name and hometown the martyr came from, so that any repercussions or reactions to the terrorist event can be channeled toward another family or town. In some cases, the cover story is used to direct government counterterrorism toward the wrong target (especially if the martyr's family is well known and the town is small). In other cases, it is used to give the impression that dozens of martyrs are coming from the same town, when in fact they are not.

In all fairness, it should be said that most militant religious groups adopt terrorism only as a tactic of last resort. The doctrine of just war has not been discussed here, but ethics and/or fair play are integral parts of most religions, and there are sometimes unwritten rules for when the "cosmic struggle"—as Juergensmeyer (2001) calls it—spills over into political struggle. Religious terrorists demonstrate marvelous ingenuity in means, methods, and timing, but their targeting is flawed, and one can only wonder how strategically effective their "symbolic" success is at times. Sometimes the whole reason for behavior is to bolster their reputation among some competing religious groups. Copycat or competitive terrorism is not beyond them, and is supported by the fact that some terrorist acts are scheduled on dates specifically designed to desecrate a competitor's religious holidays and sacred moments.

THE ECONOMICS THEORY OF RATIONAL CHOICE AS A THEORY OF TERRORISM

The discipline of economics has many concepts that are relevant to an understanding of terrorism: supply and demand, costs and benefits, and so on. However, fully developed economic or econometric models of terrorism are quite rare and often involve such things as "psychic" costs and benefits (Nyatepe-Coo, 2004). More down-to-earth economic theories can be found in the literature on deterrence. Rational choice theory, in particular, has found a place in criminology, and holds that people will engage in crime after weighing the costs and benefits of their actions to arrive at a rational choice about motivation after perceiving that the chances of gain outweigh any possible punishment or loss. Criminals come to believe that their actions will be beneficial—to themselves, their community, or society—and they come to see that crime pays, or is at least a risk-free way to better their situation. Perhaps the best-known version of this idea in criminology is routine activities theory (Cohen & Felson, 1979), which postulates that three conditions must be present in order for a successful crime to occur: (1) suitable targets or victims who put themselves at risk; (2) the absence of capable guardians or police presence; and (3) motivated offenders among a pool of the unemployed and alienated. Other rational choice theories exist that delve further into models of decision making. In the few models of collective violence that have found their way into criminology, the Olson hypothesis suggests that participants in revolutionary violence predicate their behavior on a rational cost–benefit calculus to pursue the best course of action given their social circumstances.

Rational choice theory, in political science, follows a similar line and holds that people can be collectively rational—even when making what appear to be irrational decisions for them as individuals—after perceiving that their participation is important and their personal contribution to the public good outweighs any concerns they may have for the "free rider" problem (Muller & Opp, 1986). For those unfamiliar with it, the "free rider" problem is a classic paradox in social science and economics that asks why anybody should do something for the public good, when most likely someone else will get credit for it and most everybody else will benefit merely by sitting idly by and doing nothing. Perhaps the most eloquent spokesperson for rational choice ideas in the field of terrorism is Martha Crenshaw (1998), whose writings inform the remarks that immediately follow.

Take, for example, a typical terrorist event that involves hostage-taking and the all-too-frequent hostage-killing. From an individualist rational point of view, the best choice would be to keep at least some of the hostages alive in order to bargain with the government for leniency. Yet often a collectivist rational mentality sets in, and the group choice (or group think) is to kill all the hostages. Is this killing senseless, the product of deranged minds, or an example of mob behavior? The answer is no on all counts from a rational choice point of view. It may be a reasonable and calculated response to circumstances. It may involve a collective judgment about the most efficient course of action that has the most lasting impact on observers (for social learning purposes). And most importantly, the senselessness of it all may be just what the group needs in order to make their ideological point that they are terrorists, not ordinary criminals.

Terrorism, from the rational choice point of view, is not a pathological phenomenon. The resort to terrorism is not an aberration. The central focus of study is on why some groups find terrorism useful, and, in standard control theory fashion, why other groups do not find terrorism useful. (Control theory here refers to a criminological approach that studies self-control or the inhibitions brought on by social control.) Some groups may continue to work with established patterns of dissident action. Other groups may resort to terrorism because they have tried other alternatives. Still other groups may choose terrorism as an early choice because they have learned from the experiences of others, usually through the news media. Crenshaw (1998) calls this the contagion effect and claims it has distinctive patterns similar to the copycat effect, as in other theories of collective violence (Gurr, 1970). There may be circumstances in which the terrorist group wants to publicize its cause to the world, a process Crenshaw (1995) calls the globalization of civil war.

Factors that influence the rational choice of terrorism include place, size, time, and the climate of international opinion. A terrorist plot in a democratic society is less likely to involve senseless violence than would a scheme hatched under an authoritarian regime because under the latter, with the expected repercussions, terrorists realize they have nothing to lose. Size is important because a small, elite group is more likely to resort to terrorism when the population is passive. This means that more senseless acts of violence may occur in a stable society than in one on the verge

of collapse. Time constraints are important because the terrorist group may be competing with other groups or attempting to manage a tit-for-tat strategy with counterterrorism. The climate of international opinion, if low for the problems of the host country, may force the terrorists to take actions that risk a repressive counterterrorist reaction, in hopes that their suffering will capture public attention. In short, rational choice terrorism often attempts to manage the political agenda on a world stage, and in this sense is a whole lot more than self-interested behavior.

THE GLOBALIZATION THEORY OF TERRORISM

Nassar (2004) has probably written the most interesting piece on globalization theory as it relates to terrorism, and, although his ideas are fairly critical of the Western world for exporting "nightmares" as well as dreams, he does provide a robust introduction to the complex topic of globalization. Globalization contributes to dreams, fantasies, and rising expectations, but at the same time, it leads to dashed hopes, broken dreams, and unfulfilled goals. Terrorism breeds in the gap between expectations and achievements (an idea that also constitutes a central notion in criminological strain theory). The explanation is very similar as well to what is sometimes referred to as rising expectations theory (Davies, 1962). The difference is that modern globalization theory usually adds a rich–poor dichotomy. Rich people (or nations) are seen as wanting power and wealth, and poor people (or nations) are seen as wanting justice. The rich people are seen as part of the causal factor or root cause of terrorism because they contribute to conditions that give rise to it. Perpetrators of "terrorism" (always treated as an ill-defined concept in globalization theory) are never seen as born or raised with any specific predispositions, just as oppressed and disgruntled poor people are never given the chance to find any peaceful means for achieving justice.

Globalization theory is further tied into ideas about colonialism, imperialism, and neocolonialism. The first two words often are used interchangeably to describe a set of conditions (technically, extensions of sovereignty) where the mores, values, and beliefs of the colonizers are considered superior to those of the colonized. However, when the assumption is made that (imperial) power is a necessary thing to maintain cultural superiority, this is the usual meaning of imperialism. Neocolonialism is a concept developed by Marxists and holders of certain conspiracy theories to refer to allegations about First World nations, international organizations, and/or multinational corporations employing economic, financial, and trade policies to dominate less powerful countries. Onwudiwe (2001) is typical of the criminological approach taken when globalization ideas are incorporated into theories about terrorism, and it is probably safe to say that the approach remains relatively underdeveloped but is of immense interest to some.

SOCIOLOGICAL AND PSYCHOLOGICAL THEORIES OF TERRORISM

Many contemporary sociological perspectives are dominated by approaches that focus upon the social constructions (effects) of fear or shock, and how institutions and processes—especially the media, powerful interest groups, politicians, and primary and secondary groups—help to maintain such constructions. Social constructivism is best understood as a way to look at terrorism from a relativistic standpoint (Gibbs, 1989), and such a standpoint often entails more social critique than theoretical explanation. For instance, social constructivism self-critically looks at the way terrorism can be conceived of as mythology, dialogue, and a form of communication. Social constructivism is all about consequences, not causes, and sometimes the consequence can be a cause or have an amplifying effect, such as when putting a delinquent label on a minor makes their behavior worse. Labeling theory in criminology, for example, is a social constructivist approach that goes about connecting consequences with causes in a way that is, arguably, similar but less systematic than the way sociological functionalists did it a long time ago. Traditional functionalism looks at how the consequences of living under terror affects a society's evolution with regard to values such as achievement, competition, and individualism (O'Connor, 1994). Social constructivism, by contrast, does not seem much concerned about values or evolution, which may quite possibly lead to some confusion between the *effects* of terrorism and the *effectiveness* of terrorism. It is clearly observable, nonetheless, that some societies become "softer" targets after significant terrorist events (usually because of disaster fatigue), and other societies become stronger in the long term (either through reticence or resolve). Effects vary on the basis of common sociological factors such as interaction patterns, social traits such as gender and race, and how stable are the social institutions that people form. Sociologically, terrorism is a synergistic phenomenon with causes that interact with its effects (Ross, 1999).

Many contemporary sociological theories contain strong elements of explanation at the psychological, or micro, level. One is likely to encounter the following, or some variation of them, as the five (5) main sociological theories of terrorism:

- The frustration-aggression hypothesis
- The relative deprivation hypothesis
- The negative identity hypothesis
- The narcissistic rage hypothesis
- The moral disengagement hypothesis

The frustration-aggression hypothesis is the idea that almost every frustration leads to some form of aggression, and that every aggressive act relieves that frustration to some extent. The basic notion is that stress and hassles build up until they reach a point that "breaks the camel's back," and the displacement of released energy provides some benefit in terms of catharsis or ventilation. There are different variants of this model to be found, and many forms of it exist in a variety of criminological thought.

The relative deprivation hypothesis is the idea that as a person goes about choosing their values and interests, they compare what they have and do not have, as well as what they want or don't want, with real or imaginary others. The person then usually perceives a discrepancy between what is possible for them and what is possible for others, and reacts to this discrepancy with anger or an inflamed sense of injustice. Debates exist within criminology regarding relative deprivation and terrorism, on the one hand, with the anomie or strain tradition, which finds causal influence in such objective factors as gross domestic product, and on the other hand, with the left realist tradition, which finds causal influence in subjective experiences of deprivation or discomfort.

The negative identity hypothesis is the idea that, for whatever reason, a person develops a vindictive and covert rejection of the roles and statuses laid out for them by their family, community, or society. For example, a child raised in a well-to-do family may secretly sabotage every effort made to hand them the good life on a "silver platter," deliberately screwing up in school, at work, and everyplace else until the day comes, with some apparent life-altering experience (for example, engaging in terrorism), that the long-nurtured negative identity comes out, and the subject can then make it feel more like a total identity transformation. There are many varieties of this idea that exist in a number of theories across many fields of study.

The narcissistic rage hypothesis is an umbrella idea for all the numerous things that can go wrong in child-rearing, such as too much smothering, too little smothering, ineffective discipline, overly stringent discipline, psychological trauma, coming from a broken home, and so on—all of which leads to the same effect of a "What about me?" reaction in the child. It is actually a two-way process, with the child contributing as much as the parents and other role models. This results in a damaged self-concept, a tendency to blame others for one's inadequacies, and the well-known "splitting" of self into a "good me" and a "bad me," which often forms the basis for personality disorders involving a lack of empathy for the suffering of others. There is not all that much consensus on the primal importance of narcissism, and the literature on child-rearing is full of mixed empirical results.

The moral disengagement hypothesis is the idea that encompasses all the ways a person neutralizes or removes any inhibitions they have about committing acts of horrific violence. Some common patterns include imagining oneself as a hero, portraying oneself as a functionary, minimizing the harm done, dehumanizing the victim, or insulating oneself in routine activities. Organized crime figures, for example, usually hide behind family activities with their wives and children. In the study of terrorism, there are numerous other ways that violence is rationalized that go far beyond denigrating one's enemies and beefing up oneself as a crusader (see Hacker, 1996). Terrorist rationalizations usually involve a complete shift in the way government and civil society are perceived.

Psychological approaches alone, with few exceptions (Ross, 1996; 1999), are decidedly clinical in what are often futile attempts to find something pathological in the terrorist personality. Merari (1990) provides a good overview of psychological approaches, but one of the major names in this area is David Long, former assistant director of the U.S. State Department's Office of the Coordinator for Counterterrorism, who has gone on record saying there is no such thing as a terrorist personality—but then he has said they typically suffer from low self-esteem, are attracted to groups with charismatic leaders, and enjoy risk-taking (Long, 1990). A sampling of psychological factors that have been investigated include ineffective parenting or rebellion against one's parents, a pathological need for absolutism, and a variety of other "syndromes" and hypotheses (see Margolin 1977), but study after study over the past 30 years has yielded very little valid and reliable information about the psychology of terrorists other than the following generalization:

> As far as we know, most terrorists feel that they are doing nothing wrong when they kill and injure people. They seem to share a feature of the psychological condition known as antisocial personality disorder or psychopathic personality disorder, which is reflected by an absence of empathy for the suffering of others. However, they do not appear unstable or mentally ill for this. A common feature is a type of thinking such as "I am good and right. You are bad and wrong." It is a very polarized thinking which allows them to distance themselves from opponents and makes it easier for them to kill people. It is not the same kind of simplistic thinking one would expect from someone with low intelligence or moral development. Most terrorists are of above average intelligence and have sophisticated ethical and moral development. A closed-minded certainty is a common feature of terrorist thinking (Merari, 1990).

Although what is not known about the psychology of terrorism may be more than what is known, there have been several promising attempts to merge or combine psychology with sociology (and criminal justice) into what might be called terrorist profiling (Russell & Miller, 1977; Bell, 1982; Galvin, 1983; Strentz, 1988; Hudson, 1999). This line of inquiry has a long history and includes what rare studies exist of female terrorists. The earliest study (Russell & Miller, 1977) found that the following people tend to join terrorist organizations:

- 22 to 25 years of age
- 80% male, with women in support roles
- 75% to 80% single
- 66% middle- or upper-class background
- 66% some college or graduate work
- 42% previous participation in working-class advocacy groups
- 17% unemployed
- 18% strong religious beliefs

These data, as well as other known characteristics and attributes about terrorists, have found their way into databases, some public, some private. One of the best-known databases used by researchers is the *RAND–St. Andrews Chronology of International Terrorism*. When suicide bombings became popular, Merari (1990) conducted rare interviews with terrorists and found that most suicide terrorists are between the ages of 16 and 28. Most are male, but 15% are female, and that proportion is rising. Many come from poor backgrounds and have limited education, but some have university degrees and come from wealthy families. Strength of religious belief also has increased substantially since the post-1979 rise in Islamic fundamentalism.

Criminology is quite heavily informed by sociology and psychology. What sociological and psychological approaches basically tell us is that individuals join terrorist organizations in order to commit acts of terrorism, and that this process is the same as when individuals join criminal subcultures in order to commit acts of crime. There appears to be no unique terrorist personality. Instead, there appear to be some unique subcultural phenomena that develop, support, and enhance a penchant for cold-blooded, calculated violence that, if not satisfied within a terrorist organization, might be fulfilled someplace else. Terrorism is a social activity. Individuals join a terrorist group usually after they have tried other forms of political involvement. The emotional links between individuals and the strength of commitment to their ideology appear to become stronger by the group living in the underground and facing adversity in the form of counterterrorism. Socialization in the underground is quite intense, and the criminologist Ferracuti (1982) has documented the "fantasy wars" that go on in the terrorist underground. An individual's identity may become tied to the group's identity, but it is just as likely that emotional relationships become as important as —if not more important than—the group's purpose. This means that the distribution of beliefs among members in a terrorist group may be uneven. There may be major differences between individual and group ideology. Ideology may not necessarily be the main component of motivation. From profiling terrorists for many years, it is believed that most of them are action-hungry practitioners, not theoreticians. This knowledge may provide new counterterrorism strategies that attempt to change individual beliefs and weaken group cohesion.

PSYCHIATRIC THEORIES OF MENTAL ILLNESS AS A THEORY OF TERRORISM

The leading proponent of the terrorist-as-mentally-ill approach is Jerrold Post (1984; 1990), who has gone on record saying that the most dangerous terrorist is likely to be a religious terrorist, and that all terrorists suffer from negative childhood experiences and a damaged sense of self. His analysis of the terrorist "mind-set" (a word that substitutes for terrorist personality, and technically means a fixed mental attitude or inclination) draws upon a view of mental illness that compels, or forces, people to commit horrible acts. It should be noted that the field of criminal justice holds that this is not the only possible view on mental illness. More "crazy" people come into contact with the law through sheer folly and foolishness than because of a compulsion that their mental illness made them have. Post (1990) makes a somewhat neo-Freudian distinction between terrorists who desire to "destroy the nation, or world, of their fathers," and those who desire to "carry on the mission, or world, of their fathers." In short, it boils down to the Oedipus complex, which includes hating your father, or at least the "world" he represents. There is actually some empirical support for this viewpoint. For example, when solitary terrorist offenders are studied, such as skyjackers and mail bombers, a severely dysfunctional relationship with their father is often found. The "anarchic-ideologue" terrorist, according to Post (1984), is rebelling against his or her father, and according to Kaplan (1981), has a pathological need to pursue

absolute ends because of their damaged sense of self-worth. For a review of these and related theories, see Ruby (2002).

Another analysis is provided by Jessica Stern (1999), who attempted to gain psychological insight into the distinction between "doomsday" terrorists, who would use weapons of mass destruction (WMD) that might end all life on earth, and "dangerous" terrorists, who would limit themselves to the conventional arsenal of terrorism. Stern's concern is primarily with the proliferation of WMD, and cases studies of groups such as Aum Shinrikyo, the Tamil Tigers, Al Qaeda, and Hizballah show that terrorists most likely to use WMD tend to be afflicted with paranoia and megalomania. Of these two illnesses, megalomania is more severe, and paranoia is at such a moderate level that it enhances intelligence and keeps people from becoming schizophrenics or sociopaths. Stern (1999) takes exception with arguments that terrorists suffer from any antisocial, psychopathic, or sociopathic disorder. Likewise, Victoroff (2005) says even though there is no such thing as a terrorism questionnaire (no one has validated a terrorism quotient or found a terrorist gene), it might be promising to measure their sense of oppression and feeling of subjugation, but one would have to account for the deep levels of fervor, hatred, bravado, and other psychodynamic pressures at work.

Walter Laqueur (1999) has offered the idea that the need exists to distinguish between terrorists who are "fanatics" and those who are "extremists." The standard meaning of these terms is that fanatics are religious zealots, and extremists are political zealots, but Laqueur (1999) strips away any religious connotation, and says that most terrorists are fanatics. The concept of fanaticism carries some implications of mental illness, but is not in itself a diagnostic category. Laqueur (1999) claims that fanaticism is characterized by excessive cruelty and sadism. Others (Taylor 1991) have pointed out that fanaticism is characterized by the following:

- Prejudice toward out-groups
- Authoritarianism
- An unwillingness to compromise
- A disdain for other alternative views
- A tendency to see things in black and white
- A rigidity of belief
- A perception of the world that reflects a closed mind

To this, one might add the concept of Machiavellianism (Oots & Wiegele, 1985), which refers to an extreme form of the psychological trait of manipulativeness. Terrorists are disposed to manipulate not only their victims, but their audience as well. Both the timing of an event and the aftermath of an event are manipulated by terrorists. For example, the counterterrorism reaction by authorities is manipulated. The press and public are manipulated, with terrorists doing everything they can to work the media and obtain liberal press coverage. The fact that terrorism is aimed more at the audience than at the victim has provided numerous points of conjecture for criminologists. It has been the starting point for labeling and constructivist approaches. It has been the source of most theoretical models of terrorist contagion (whereby different terrorist groups compete with one another for media attention) as well as most theoretical models of copycat behavior (whereby different terrorist groups try to outdo a previous group with the harm inflicted). What is not widely known is that the almost addictive, or at least cyclic, quality of a need to manipulate audience reaction has inspired those with a biological or physiological interest.

BIOLOGICAL AND PHYSIOLOGICAL THEORIES OF TERRORISM

David Hubbard (1983) was one of the first biological researchers of terrorism, and his line of work is similar to the familiar cycle of violence hypothesis in criminal justice. In this view, people who commit repetitive and cyclical acts of violence (which would include wife beaters, rapists, and serial killers) are driven by hormonal or neurochemical fluctuations in their body or brain chemistry. Three compounds, in particular, have been singled out as having abnormal levels among terrorists: norepinephrine, acetylcholine, and endorphins. Of these, norepinephrine is suspected as being the most influential because it is associated with the so-called flight or fight mechanism in human biology. The theory of "fight or flight" was developed by W. B. Cannon back in 1929 and refers to a state of arousal under stress in which the heart, lungs, and muscles operate more efficiently. As it applies to terrorism (and crime), the behavioral requirements of such activities (fighting exhilaration before an event and fleeing poorly done manipulation of an audience after an event) produce a syndrome of physiological need for arousal at fairly regular intervals. Motives for terrorism appear to be quite stable when the biological viewpoint is taken, and it is possible to link a variety of aspects in the typical terrorist profile with biological factors. For example, xenophobia may be biologically induced because the fear of strangers tends to be discernible even at infantile stages of development. Other explanations might involve insights from sociobiology (Wilson, 1975) or kin selection theory in evolutionary psychology (Daly & Wilson, 1994). For more information on biological factors as the cause of criminal behavior, see almost any criminology textbook.

TRADITIONAL CRIMINOLOGICAL THEORIES APPLIED TO TERRORISM

It is not easy applying traditional criminological theories to terrorism. Most of these theories were designed to explain ordinary street crime, such as robbery or burglary, and have a certain hardiness to their perspectives that makes them difficult to extend. Ruggiero (2005) is typical of those who have attempted to apply such theories or suggest various extensions, starting with Durkheim's functionalism by asking whether Durkheim would see terrorism as part of the "normality of crime" or as a clearly unacceptable, dysfunctional form of crime. On the one hand, Durkheim said that all crime serves positive functions (of innovation and evolution), but on the other hand, the organic metaphor that Durkheim used seems to suggest that some forms of crime only cause disintegration and are cancerous. The Chicago school of disorganization in criminology would presumably focus on the distinctiveness of different social worlds between terrorists and nonterrorists, analyzing the communication blockages, for example. Strain theorists would likely argue that terrorism is inevitable as a manifestation of the broken promise that everybody can rise from rags to riches, and study the adaptations Merton described as innovation or rebellion. Learning theorists would likely emphasize the importance of role models or the "techniques of neutralization" involved along with the drift into a terrorist lifestyle. Labeling theorists would probably say, cynically but truly, that terrorism is "what the other person does." Control theorists would likely focus on terrorists being unattached, unloved, uncommitted to education or business, uninvolved in conventional tasks, and having their hands so idle that time becomes the "devil's playground" for them. Conflict theorists would probably focus on the presence or absence of associations that provide room for collective action and permanent confrontation, although more radical versions of conflict theory might glamorize terrorism as proto-revolutionary action. Integrated theories would likely focus on the influences of aggressive proneness, provocation, and the support of third parties.

THEORIES UNIQUE TO DOMESTIC TERRORISM

Freilich (2003) does a good job of reviewing the theories in this category, which is a relatively small area of research that tends to be studied within a field called the sociology of social movements. There are mainly three groups of theories. The first is called economic/social integration theory, and it holds that high concentrations of farming, economic depression, and social disorganization are all related to high levels of domestic terrorist activity and, in particular, militia movements. In some varieties, it tends to be a kind of "farm crisis" or "agrarian reform" theory frequently used by those who study underdeveloped nations. The second group is called resource mobilization theory, and it suggests that states that are more prosperous and socially integrated would tend to develop more domestic terrorist activity, on the basis that group competition for power and resources is more intense. The third group of theories is called cultural theories, which propose that states experiencing greater cultural diversity and female empowerment along with increasing paramilitarism are likely to develop greater levels of domestic terrorist activity. In terms of research findings, more empirical support seems to exist for the third set of theories, at least according to Freilich (2003), although resource mobilization theory tends to dominate the theoretical literature. See Jenkins (1983) for the definitive list of resources, such as money, organizational facilities, manpower, means of communication, legitimacy, loyalty, authority, moral commitment, and solidarity. In general, there is more empirical support for the idea that domestic terrorism more often plagues richer and affluent nations than it does poor ones.

Briefly, resource mobilization theory describes the process by which a group assembles material and nonmaterial resources and places them under collective control for the explicit purpose of pursuing a group's interests through collective action. Collecting resources must be accompanied by mobilization of resources. A group may prosper yet still not contend for power. Four central factors condition the process of mobilization: organization, leadership, political opportunity, and the nature of political institutions. Strong horizontal links between members of a group provide the best organizational structure. Leaders who make themselves available to members and take an interest in members' grievances tend to make the best leaders. Political opportunities refer to moments when the "time is ripe" for action, and groups that seize upon such opportunities tend to succeed. Political institutions refer to moments when the existing political parties are weak or fractured, and these are seen as times when domestic terrorist groups will best succeed by taking action.

CONCLUSIONS

Although the theoretical and conceptual work undertaken by criminologists on terrorism may seem vast and extensive, the fact of the matter is that much of it is speculative and only in the beginning stages of model development. Clearly there are difficulties in pinpointing the key causal variables, and much of this is due to the nature of criminology itself borrowing from different disciplines, so much so that competitive explanations run the risk of canceling one another out. For example, it may be that labeling and constructivist explanations that focus upon identity salience (Arena & Arrigo, 2006) offer a better way to address the cyclic nature of terrorist behavior than do psychological or biological explanations. Sociological theories that avoid psychological reductionism tend to be heavily favored, and there are those (Horgan, 2003) who strongly protest the absurdity of searching for the so-called terrorist personality. One may at least take comfort in criminology's provision of multiple frameworks enabling researchers room to explore and triangulate their efforts. Theory-driven research is better than dredging around in the data, hoping for an idea to pop up. The criminology of terrorism also may provide tentative implications and applications for the law enforcement, military, and intelligence communities. Theoretical work can contribute not only to a more informed accounting and understanding of terrorism; it also can assist in policy formation by forestalling a tendency to reinvent the wheel whenever the next disastrous and perplexing incident takes place.

REFERENCES

Arena, M. & Arrigo, B. (2006). *The terrorist identity.* New York: NYU Press.

Bell, B. (1982). "Psychology of leaders of terrorist groups." *International journal of group tensions* 12: 84–104.

Bocock, R. (1986). *Hegemony.* London: Tavistock.

Cohen, L. & Felson, M. (1979). "Social change and crime rate trends: A routine activities approach." *American sociological review* 44: 588–608.

Crenshaw, M. (Ed.) (1995). *Terrorism in context.* University Park: Pennsylvania State University Press.

Crenshaw, M. (1998). "The logic of terrorism: Terrorist behavior as a product of strategic choice." In W. Reich (Ed.), *Origins of terrorism.* New York: Woodrow Wilson Center Press.

Daly, M. & Wilson, M. (1994). "The evolutionary psychology of male violence." In J. Archer (Ed.), *Male violence* (p. 253–288). London: Routledge Kegan Paul.

Davies, J. (1962). "Towards a theory of revolution." *American sociological review* 27: 5–18.

Ferracuti, F. (1982). "A sociopsychiatric interpretation of terrorism." *Annals of the American academy of political and social science* 463: 129–141.

Ferrell, J. (1997). "Against the law: Anarchist criminology." In B. MacLean & D. Milovanovic (Eds.), *Thinking critically about crime.* Richmond, BC: Collective Press.

Ferrell, J. (1999). "Anarchist criminology and social justice." In B. Arrigo (Ed.), *Social justice/Criminal justice* (p. 91–108). Belmont, CA: Wadsworth.

Freilich, J. (2003). *American militias: State-level variations in militia activities.* New York: LFB Press.

Galvin, D. (1983). "The female terrorist: A socio-psychological perspective." *Behavioral sciences and the law* 1: 19–32.

Georges-Abeyie, D. & Hass, L. (1982). "Propaganda by deed: Defining terrorism." *Justice reporter* 2: 1–7.

Gibbs, J. (1989). "Conceptualizations of terrorism." *American sociological review* 53(4): 329–340.

Gurr, T. (1970). *Why men rebel.* Princeton, NJ: Princeton University Press.

Hacker, F. (1996). *Crusaders, criminals, crazies: Terror and terrorists in our time.* New York: Norton.

Hoffman, B. (1993). *Holy terror.* Santa Monica, CA: RAND.

Horgan, J. (2003). "The search for the terrorist personality." In A. Silke (Ed.), *Terrorists, victims, and society* (p. 3–27). West Sussex, UK: Wiley.

Hubbard, D. (1983). "The psychodynamics of terrorism." In Y. Alexander et al. (Eds.) *International violence* (pp. 45–53). New York: Praeger.

Hudson, R. (1999). *Who becomes a terrorist and why.* Guilford, CT: Lyons Press.

Jenkins, J. (1983). "Resource mobilization theory and the study of social movements." *Annual review of sociology* 9: 527–553.

Juergensmeyer, M. (2001). *Terror in the mind of God: The global rise of religious violence.* Berkeley: University of California Press.

Kaplan, A. (1981). "The psychodynamics of terrorism." In Y. Alexander & J. Gleason (Eds.), *Behavioral and quantitative perspectives on terrorism* (p. 35–50). New York: Pergamon.

Kraemer, E. (2004). "A philosopher looks at terrorism." In A. Nyatepe-Coo & D. Zeisler-Vralsted (Eds.), *Understanding terrorism* (p. 113–131). Upper Saddle River, NJ: Prentice Hall.

Kushner, H. (2003). *Encyclopedia of terrorism.* Thousand Oaks, CA: Sage.

Laqueur, W. (1999). *The new terrorism.* New York: Oxford University Press.

Long, D. (1990). *The anatomy of terrorism.* New York: Free Press.

Margolin, J. (1977). "Psychological perspectives in terrorism." In Y. Alexander & S. Finger (Eds.), *Terrorism: Interdisciplinary perspectives.* New York: John Jay Press.

Merari, A. (1990). "The readiness to kill and die: Suicidal terrorism in the Middle East." In W. Reich (Ed.), *Origins of terrorism.* Cambridge, UK: Cambridge University Press.

Muller, E. & Opp, K. (1986). "Rational choice and rebellious collective action." *American political science review* 80: 471–487.

Nassar, J. (2004). *Globalization and terrorism.* Lanham, MD: Rowman & Littlefield.

Nyatepe-Coo, A. (2004). "Economic implications of terrorism." In A. Nyatepe-Coo & D. Zeisler-Vralsted (Eds.), *Understanding terrorism* (p. 77–89). Upper Saddle River, NJ: Prentice Hall.

O'Connor, T. (1994). "A neofunctional model of crime and crime control." In G. Barak (Ed.), *Varieties of criminology* (p. 143–158). Westport, CT: Greenwood Press.

Onwudiwe, I. (2001). *The globalization of terrorism.* Andershort, Hampshire: Ashgate Press.

Oots, K. & Wiegele, T. (1985). "Terrorist and victim: Psychiatric and physiological approaches." *Terrorism: An international journal* 8(1): 1–32.

Passmore, K. (2002). *Fascism: A very short introduction.* New York: Oxford University Press.

Post, J. (1984). "Notes on a psychodynamic theory of terrorist behavior." *Terrorism: An International Journal* 7: 241–256.

Post, J. (1990). "Terrorist psycho-logic: Terrorist behavior as a product of psychological forces." In W. Reich (Ed.), *Origins of terrorism* (p. 25–40). Cambridge, UK: Cambridge University Press.

Purkis, J. & Bowen, J. (Eds.) (1997). *Twenty-first century anarchism.* London: Cassell.

Ranstorp, M. (1996). "Terrorism in the name of religion." In R. Howard & R. Sawyer (Eds.), *Terrorism and counterterrorism* (p. 121–136). Guilford, CT: McGraw-Hill.

Ross, J. (1996). "A model of the psychological causes of oppositional political terrorism." *Peace and conflict: Journal of peace psychology* 2(2): 129–141.

Ross, J. (1999). "Beyond the conceptualization of terrorism: A psychological-structural model." In C. Summers & E. Mardusen (Eds.), *Collective violence.* New York: Rowen & Littlefield.

Ruby, C. (2002). "Are terrorists mentally deranged?" *Analyses of social issues and public policy* 2(1): 15–26.

Ruggiero, V. (2005). "Political violence: A criminological analysis." In M. Natarajan (Ed.), *Introduction to international criminal justice* (p. 35–41). New York: McGraw-Hill.

Russell, C. & Miller, B. (1977). "Profile of a terrorist." *Terrorism: An international journal* 1(1): 17–34.

Smith, B. (1994). *Pipe bombs and pipe dreams: Terrorism in America.* Albany: State University of New York Press.

Smith, B., Damphousse, K., Jackson, F., & Karlson, A. (2002). "The prosecution and punishment of international terrorists in federal court: 1980–1999." *Criminology and public policy* 1(3): 311–338.

Stern, J. (1999). *The ultimate terrorists.* Cambridge, MA: Harvard University Press.

Stitt, G. (2003). "The understanding of evil: A joint quest for criminology and theology." In R. Chairs & B. Chilton (Eds.), *Star Trek visions of law & justice* (p. 203–218). Dallas, TX: Adios Press.

Strentz, T. (1988). "A terrorist psychological profile." *FBI Law enforcement bulletin* 57: 11–18.

Taylor, M. (1991). *The fanatics.* London: Brassey's.

Tifft, L. (1979). "The coming redefinition of crime: An anarchist perspective." *Social problems* 26: 392–402.

Victoroff, J. (2005). "The mind of a terrorist: A review and critique of psychological approaches." *Journal of conflict resolution* 49(1): 3–43.

Wilson, E. (1975). *Sociobiology: The new synthesis.* Cambridge, MA: Harvard University Press.

CHAPTER 3

Christian Identity

Kevin Borgeson and Robin Valeri

> There is a good deal of gabbling about the threat of terrorist use of nuclear, chemical, and biological "Weapons of Mass Destruction" against the West. There is no threat. It's already been detonated. It was a biological weapon. Its disease was JEWS.
>
> — Richard Butler, *Aryan Nations*

Some social scientists (Post, 1984; 1990) see religious-based terrorism as one the biggest threats to society, due to justifications based on a "higher power." For those in the Aryan Nations and the Ku Klux Klan, religious justification comes from a religious doctrine they refer to as "Two Seed Line Christian Identity," which will be outlined in this chapter. Christian Identity does not use a historical analysis of the Bible; it must be read as a revisionist interpretation that proves their racist and anti-Semitic dogma. In James Aho's *The Politics of Righteousness*, he points out that Christian Identity (CI) adherents "see the world through three specific doctrines: 'dualism, conspiracy, and the coming apocalypse'" (Aho, 1990, p. 220). Christian Identity inherited these features from another religious phenomenon dating back to the late 1880s, British Israelism. "British Israelism (BI), in the most general terms, refers to the belief that the British are lineal descendants of the 'ten lost tribes' of Israel" (Barkun, 1997, p. 4). British Israelism had its earliest advocate in the late 1700s in Richard Brothers. According to Michael Barkun, the author of *Religion and the Racist Right: The Origins of the Christian Identity Movement*, Brothers began having visions in 1791 that he was to help the Jews. Around 1793:

> . . .he concluded that he had a divine mission to lead the Jews back to Palestine. He decided that he himself was a descendant of the house of David, and that most Jews were hidden between Existing European, and particularly British Peoples, unaware of their exalted biblical lineage. This idea of "hidden Israel" that believed itself Gentile, ignorant of its true biological origins, marks the initial appearance of what is to become British–Israleism's "central motif" (Barkun, 1997, p. 6).

Brothers did not garner much support, and was eventually institutionalized.

One of the earliest leaders to gain support for British Israelism was John Wilson. Unlike Brothers, Wilson believed that the Jews did not have a special covenant with God. Wilson did not believe that there was such a place as Israel, and the biblical Scriptures told him that Israel was a people, not a place. Wilson was crucial in today's Christian Identity movement because he introduced the Idea that the Anglo-Saxon, Caucasian race was the true house of Israel. He believed that the special ordinances given to Israel were not given to the Jews but to the white Caucasian race. Unlike preceding BI believers—such as Richard Brothers, who felt Jews were equal and had a place in the New Jerusalem—Wilson felt that Jews did not hold a "religious status equal to their newly discovered Northern European Brethren" (Barkun, 1997, p. 7). Wilson believed that the Caucasian race could be traced back to the English throne, and that England would be the place for the "new Jerusalem."

Probably the best-known British Israelism preacher of his time was Edward Hines, who taught BI in the late 1880s. Hines was influenced by Wilson, but took a different look at the emphasis of Britain in biblical prophecy. The Hines philosophy of BI played on nationalistic and patriotic themes that Wilson's type of BI excluded. Michael Barkun contends that the Hines BI style played into the sentiments of England at that time (1870s). Germany was seen as a rival among the Brits, and a xenophobic philosophy permeated the larger social system in relation to the Germanic people. The British began seeing the Germans as a threat, and did not want to share the special covenant of biblical prophecy with them. Like most religions, BI developed several denominations and belief systems in regard to Jews, Jerusalem, the lost tribes, and the oncoming apocalyptic end to society.

One of the early denominations was pyramidology, the belief that the truth to the end of the world is locked in the structure of the pyramids. This eventually extended to beliefs in numerology—in which most answers could be found (lost tribes, etc.) in significant dates and their sequencing. For instance, Aho (1990, p. 11) points out the triangulation method used by various BI, in order to discover the locations of the lost tribe:

> The five planets visible to the naked eye plus the sun and the moon multiplied by 360=2,520 years. This we are told is seven prophetic cycles. But 2,520 years added onto 576 B.C., the year when Benjamin was captured Babylonia, is exactly 1945, the year when Iceland became independent from Denmark. This is "conclusive evidence" that Iceland was the final stopping place of Benjamin.

Although this phenomenon did gather support, it had one element that would eventually become its undoing: predictions. Because BI members felt that all social events could be predicted by using the "power of the pyramids," including the oncoming apocalypse, they gave exact dates of the end of the world. When the end did not happen, leaders had to explain away what happened. Most followers became disillusioned by these inaccurate predictions, turning to CI, which did not rely on specific dates, but just that "the end times are near."

It was not just inaccurate predictions that led to the acceptance of CI into the American society. In the early 1920s, BI started to make a stance that would resonate with CI adherents in America: anti-Semitism. The most famous case was that of William J. Cameron, the editor of the *Dearborn Independent*, owned by Henry Ford. From 1920 to 1927, Cameron ran a series of articles titled the "Jewish Problem" (Ridgeway, 1995). Also, at this time, the Ku Klux Klan had a premier status in the United States. The KKK influenced the larger social order, and BI begins to incorporate tenets that would become CI.

As BI died down, CI began to emerge in its place. The most popular denomination among CI adherents is the "two seed theory." Adherents of this denomination believe that the Old Testament and the New Testament need to be looked at together. The remainder of this chapter will outline the eight most important elements of CI: pre-Adamite, Serpent seed,

Noah, Babel, Abraham, Jacob / Esau, Jesus, and Armageddon. The biblical Scriptures that they use to justify their hatred toward those of the Jewish faith will also be addressed.

PRE-ADAMIC MAN

> God blessed them, and God said unto them, be fruitful, and multiply, and replenish the earth, and subdue it: and have dominion over the fish of the sea, and over the fowl of the air, and over every living thing that moveth upon the earth (Gen. 1:28).

According to CI believers, you have to believe that not all men and women are created equal. This is exemplified by the following quote by this Aryan Nations leader:

> This is what I tell people when they ask me if all people were created equal. I say no. In Genesis 1:31 God said that everything that he created was very good. Well if he says that it is very good who am I to say that it is not? But, they were created differently. You can't make a dog meow or a cat bark, you simply can't make something out of somebody that they weren't created to be. There is no way on God's green earth that any other race will be equal to the white race. This is not said to be mean, this is a statistical fact that the white race is the most intelligent race on the planet. All you have to do is look around you. Everywhere you see a high rate of civilization you find the white man.[1]

To proponents of this religion, God chose the Caucasian people of the world as his chosen people, not the Jews. They believe the Bible was written for whites, and for no one else. Whites are the true house of Israel, and all of God's prophecies are for those of Anglo-Saxon, Germanic, and Scandinavian heritage. Most Identity believers feel that the world was created in eons, not days. Over this long period of time (millions of years), God produced a series of creations in succession to provide the living components needed for each phase. In order to understand Christian Identity, we have to begin with what they believe was the separate creation

1. Unless otherwise specified, all quoted material in this chapter stems from interviews conducted with members of the Aryan Nations, all of whom were guaranteed anonymity. The interviews were conducted between August 2001 and December 2005.

of minorities—or what some Identity supporters refer to as "mud people," or what is called "pre-Adamic" man by some Identity evangelists.

Minorities, according to Identity, are to have a subservient presence on the earth to the "Adamic pure white race." In Genesis 1:26, it states: "Let us make man in our image, after our image and let them have dominion over the fish of the sea and the fowl over the air, and over the cattle and over every creeping thing that creepeth upon the earth." And therefore, in Genesis 1:27, CI adherents believe that minorities were created—not as equals, but to be used by whites. Identity believers call this "the sixth day." This is the time that all subservient beings were made. One Aryan Nations member described it this way: "Those pre Adam and Eve creations are placed here for some purpose, and that purpose is to serve the white Adamic race." For most CI advocates, God justifies the term "white supremacy." According to this member, they have been wrongly accused by society; they are not racists; they are doing the work of God:

> If a black holds up his fist and says black power, nobody says anything. If a Mexican holds up his fist and says viva la rosé, nobody says anything. But let a white man be proud of his race, or be concerned about it, now he's a racist. They call us white supremacists; well I can show you several places where God said we are a special people, a peculiar people, to be above all the other peoples of the earth. If that sounds like white supremacy to you, take it up with God he said it, I didn't.

In the Identity ministry, the semantic distinction of male and female becomes important. For Identity enthusiasts, male and female refers to the creations (pre-Adamites), and man and woman refer to Adam and Eve. This is important because in Genesis 3:15, God refers to the damnation of the satanic seed line and places "enmity" between the seed of the serpent and the seed of the woman.

How Identity members know that pre-Adamites existed is in the content of the passage of Genesis 1:28, where it was stated that "God blessed them, and God said unto them, 'Be fruitful, and multiply, and replenish the earth, and subdue it: and have dominion over the fish of the sea, and over the fowl of the air, and over every living thing that moveth upon the earth.'" To Identity adherents, this is the proof that all human beings are not equal, and were not created at the same time because, according to one Aryan Nations member, "in order to replenish something, there needs to be an existence prior to the replenishing." Because Identity believers advocate white supremacy, it is only logical that the next step for them would be to claim that whites are given a special existence, through Adam, and that they have a special quality. That quality is a soul.

This is seen in the verse where God breathed life only into Adam and his offspring. In other words, what this means to CI believers is that minorities (pre-Adamites) do not have a soul. Only the true Adamic (white) race was given a soul.

> And the LORD God formed man [of] the dust of the ground, and breathed into his nostrils the breath of life; and man became a living soul (Gen. 2:7).

One of the most important passages in understanding CI white supremacy is that of the creation of Adam and Eve, and what CI adherents refer to as "the breath of life." Most Identity believers accept as the truth that God gave the whites a spiritual uniqueness (i.e., a soul) that is destined upon them after conception. If God believes that there is "even one ounce of bad blood (i.e., race mixing), then that individual will not have a place in heaven."

The creation of Adam came after God had all the right sequences in place, on the "eighth day." In all the days preceding, "[God] brought the herbs, the grass, the trees, the fish, and cattle to prepare for [Adam and Eve] because he was supplanting life giving standards" (Wickstrom, n.d.). Adam was created "to live in innocence, and with full, unimpaired, responsibility because there was no evil nature within him. Both were created with a bias toward good and were given freedom of the will" (Gayman, n.d., p. 58). According to Identity members, minorities cannot have a soul because they "genetically are not human; they are beasts of the field." Only true white Aryans, those who come from the seed of Adam, are considered to have a soul, because they (Adamites) are the only race that can blush. "Aw Dawm" (Adam) means one who blushes and shows color in the face.

THE SERPENT SEED

> And I will put enmity between thee and thy woman, and between thy seed and her seed; it shall bruise thy head, and thou shalt bruise his heel (Gen. 3:15).

Genetic difference plays an important part in understanding Christian Identity's interpretation of the Bible. The serpent seed is portrayed as an apocalyptic battle between good and evil, and in the end (Armageddon), Yahweh (the Hebrew writing of God) will come down and ask the true white Aryan race to pick up arms and become his "battleaxe," laying the evil seed and redeeming the true white seed line on Israel soil (i.e., United States of America).

CI believers think that "the seduction of Eve was purely sexual in nature" (11th Hour Remnant Messenger). For Identity believers, "Israel is not a land, it is a people" (Wickstrom, n.d.), and those people are the Adamic white race. According to most Identity members, "The most important verse in the Bible is Genesis 3:15: 'And I will put enmity between thee and the woman, and between thy seed and her seed; it shall bruise thy head, and thou shalt bruise his heel.'" This is seen as the fight between good and evil, whites and Jews, God and the devil. Proof of two seeds, according to members, is in Genesis 3:13, which states: "And the Lord God said unto the woman, what [is] this [that] thou hast done? And the woman said, the serpent beguiled me, and I did eat." According to the 11th Hour Remnant Messenger, an anti-Jewish resource center, "The references thereafter to the covering of the genitals (Genesis 3:7) and the punishment of pain in childbirth (Genesis 3:16) could hardly be due to the literal consumption of fruit from a tree."

As one Aryan Nations member stated:

> She did not eat a literal piece of fruit. You take two key words beguiled and eat and put them back in the Hebrew. You find that one of the words for beguiled is to morally seduce. What it means for the word eat is to lay with. When you put this back into context it says that the serpent seduced me and I lay with him. You can verify this in the parabel of the wheat and the tare in Mathew 13. There were two seeds planted in the same field at the same planting time. When the blades sprung forth, so did also the tares. Cain and Abel were twins, but they had two different fathers.

Dan Gayman, the author of *The Two Seeds of Genesis 3:15*, states that the possibility of one birth producing twins by two different fathers is not far-fetched. In Gayman's book, he cites the example of Grete Bardaum, who in 1987 gave birth to fraternal twins—one black, the other white. The example is taken from a *Newsweek* article that stated that a black American G.I. was the father of the black child, and a German white man was the father of the white child.

Eve gave birth to two sons, Cain and Abel. The Bible depicts that Cain slew his brother Abel, and God declared that Cain was bred from a satanic seed line:

> Not of Cain who was of that wicked one (Satan his father), and slew his brother. And wherefore slew he him? Because his own works were evil, and his brothers righteous (11th Hour Remnant Messenger, I John 3:12 KJV).
>
> And Adam knew his wife again; and she bare a son, and called his name Seth: For God, [said she], hath appointed me another seed instead of Abel, whom Cain slew (11th Hour Remnant Messenger).

Seth, the replacement, was "to establish a true bloodline through which Messiah, Jesus the Christ, is to redeem mankind from the fall of Adam. Without this true, untainted blood line, His chosen cannot receive promise of redemption by the propitiation of Christ on the cross" (11th Hour Remnant Messenger). Cain becomes upset because God did not bless him, and he is asked to leave the kingdom of God. This is important to CI adherents because CI people see this as the introduction of the satanic seed line (Jewish) into the Adamic world. After God banished Cain to the Land of Nod, Cain started his own seed line with pre-Adamites.

It was at this point that God wanted to establish a true white seed line by placing Seth into the seed line to keep the true white Christian identity going. Due to Cain's jealousy, God took on names that resembled those of the true blessed Adamic seed line (i.e., Seth's) "to confuse the bloodline and lineage with that of Cain. It would be easy to confuse this fact in the beginning, and there was much intermingling of race and bloodline as a result of this deception" (11th Hour Remnant Messenger).

Seth Seed Line

- Enos
- Cainan
- Mahalaleel
- Jared
- Enoch
- Methesulah
- Lamech
- Noah
- Shem Japeph Ham

Cain Seed Line

- Enoch
- Irad
- Mehujael
- Methusael
- Lamech

Due to the extreme amount of race mixing that went on for generations, God became angry at his children because he preaches "Kind after their Kind," meaning that each creation is to stay among themselves: birds with birds, cattle with cattle, male with female (pre-Adamites), and man with woman (true Aryan Adamites). In order to rid the world of this evil, God established the Flood to wipe out the genetic improprieties on earth.

NOAH

> And the Lord said, I will destroy man whom I have created from the face of the earth; both man, and beast, and the creeping thing, and the fowls of the air; for it repenteth me that I have made them. But Noah found grace in the eyes of the Lord (Gen. 6:7–9).

Among Identity supporters, there are two lines of thinking regarding the Flood: that it was universal, and that it was local. Most members of the Aryan Nations believe that the Flood was not universal, but that it was restricted to the Tarim Basin, north of what is today Tibet.

> The Egyptians wrote about Noah's flood, the Chinese wrote it, and even call him Noah. How could they have written about it if they were destroyed in the flood? This is the verse that no one can get over, around, under, or through: Genesis: 9 is the last chapter regarding the flood. 10:1 they are getting off the boat and having babies. Ham had Cush, Cush had Nimrod. Verse 10 of chapter ten says Nimrod became the mighty ruler of four cities and the land of shiner. How in the world were there enough people to populate four cities? In less than two hundred years, unless there were people already there.

Because the Flood was localized, non-Adamic people outside the region were saved from this catastrophe. God was punishing only those who had had sexual relations with "six-day creation" beings and were their offspring. God was not mad at those who were "sixth day" creations and did not intermingle with those of Adamic descent. Most philosophical interpretation of CI religion is an offshoot of a biological approach, as stated by this believer:

> They say that the flood drowned everyone upon the face of the earth. Everybody that survived is a descendant of the three sons of Noah. Noah and his wife and sons Hamm, Shem, and Jacob begotten all the people on the face of the earth. That means that they were the common parents within a period of 2,400 years before Christ, of all the Asiatics,

> of all the Negroes, and all the white men on the face of the earth. It is biologically unsound, unscientific; it is not genetic, and there is no proof in it.

For the supporters of this ideology, God was trying to create a pure bloodline:

> To accomplish this purpose in redeeming Adamic man through the virgin birth through in his incarnation in Jesus Christ. The Bible tells us that all living perished except for Noah and his family. It can only reference to those that who lived in the proximity of the floods, those who are the offspring of man. No others are possible the Bible is written to and for his chosen Adamic man (11th Hour Remnant Messenger).

Noah was chosen because he resided in the Tarim Basin area and was of pure Adamic stock. Noah and his family were saved from the Flood and its wrath because of their strict adherence to the structural laws of kind after kind. God wished to create a pure bloodline for the redemption of the Adamic race, and for the resurrection of God in the flesh form of Jesus Christ. "Several million of Adamic mankind and their mixed blood offspring that lived there, all that had turned from God, would perish, as well as all the animals that were indigenous to the area that were not taken on the ark" (11th Hour Remnant Messenger). Noah and his family stayed on the Ark for more than seven months. It wasn't until an earthquake created a crack in the basin, allowing the water that had built up during the Flood to dissipate, that Noah and his family left the Ark.

The 11th Hour Remnant Messenger gives four main reasons for why they feel the Flood happened. The purpose of the Flood was to destroy:

1. The offspring from Satan's fallen angels who seduced Adamic women and produced Giant/Nephalim
2. Adamic men/women who have race mixed
3. Adamic men/women who have mixed with the offspring of Cain
4. Evil and violence of the Tarim Basin

CI advocates claim that because pre-Adamites still exist to this day, the Flood upon the earth was restricted to the basin, and because minorities exist today, they must have lived outside the area. After the Flood, Noah's sons, Ham and Japheth, were dispersed "to what is now North Africa and Eastern Russia, respectively" (11th Hour Remnant Messenger).

PREVENTIVE MEASURES

> Therefore is the name of it called Babel; because the Lord did there confound the language of all the earth: and from thence did the Lord scatter them abroad upon the face of all the earth (Gen. 11:9).

In the story of Babel, the Bible describes how Nimrod is the builder of Babel. At this time, there was only one language in the world. God saw the making of Babel as idolatry, seeing man slip back into violating his divine laws. Because there were descendants of Ham, Noah's son, God felt that he could not destroy the city. In order to add confusion to the world, God gave different languages according to their "family, tribe, or group of people" (11th Hour Remnant Messenger). Some Aryan Nations members described this to me as a preventive measure on God's part, in order to prevent any future race mixing.

> Other religions (meaning main-stream Judea Christian) will say well Hamm married a Black and Japheth married an oriental; another will say Hamm turned Black and Japheth became oriental. In Genesis 1 it says about ten times that kind goes with kind. He confused the language to create confusion among those there to prevent people from temptation.

"The earth was all one land mass at this time. It was easy for all peoples to walk or ride beasts of burden or horses all over the land mass" (11th Hour Remnant Messenger). Because there were other minority groups—blacks and Orientals—God broke apart the landmass, making what today are our continents. According to CI believers, God did this in order to make it more difficult for people to unite to do future evil deeds. For most CI supporters, Genesis 11:7 proves that God planned this separation. Some believers see it as a sign that God wished to keep the races separate. It is a rationale still used today by some Identity supporters. By using Babel, they rationalize that minorities should live on the planet, but only in separate areas, isolating themselves from the true Adamic seed line. Babel has become a justification by non-CI adherents also; most state that God separated the races for a reason, and that most of today's blacks should be sent back to Africa, where God sent them. To most CI supporters, this is not being racist; this is abiding by the laws of God.

Because the Bible was written for those of Anglo-Saxon, Germanic, and Scandinavian heritage, Identity members purport that those who are of non-Adamic races will not understand the Bible because they are not true Hebrews. This can be seen in the following translation of Romans 9:3 by a CI believer:

> To my kinsmen according to the flesh who are Israelites. He is telling you in this that his kinsmen are Israelites. To whom pertained the adoption and the glory covenants and the giving of the law and the service of God and the promises. That's everything that the Bible has to offer. Adoption, covenants, and promises and he says right there that it only belongs to Israel.

Believers feel that teachings that exist in the pulpits of today's churches are misled because they are not the entire chosen race of God. Therefore, when someone says that the Bible is being misread by Christian Identity adherents, they state that it is the pagan religions that exist in today's society that do not understand the Bible, because it was never written for them. Until the pagans wish to open their eyes and see the true words of God, they will not be Christians. According to Identity, Noah's sons departed to the land of North Africa and Eastern Russia, and the Adamic true bloodline is established through the Shem lineage.

ABRAHAM

> Now the Lord has said unto Abram, get thee out of thy country, and from thy kindred, and from thy father's house, unto a land I will show thee. And I will make of thee a great nation, and I will bless thee, and make thy name great shall be a blessing (Gen. 12:1–2).

According to the 11th Hour Remnant Messenger, this is the "the first covenant with Adamic man, now Hebrew man, since the flood in the man of Abram/Abraham, God the father is fulfilling his plan to redeem his chosen children from the bondage of Satan and his followers" (11th Hour Remnant Messenger). Abraham and his wife gave birth to Isaac, whom God also blessed. It would be Isaac's son Jacob on whom the CI movement based the true race of Israel. "Jacob was the second born to Isaac and Rebecca, a fraternal twin having Isaac as his father. The

Adamic/Hebrew seed line will remain intact through this union" (11th Hour Remnant Messenger).

The other fraternal twin born to Rebecca was Esau, Satan's child. According to some Aryan Nations members, "It is this union that is at battle to this day. And due to this union of the bad seed, God blesses Jacob and changes his name to Israel." To these believers, Israel is not a land but a race. To CI adherents, the covenants of God would go to Abraham and to his seed.

This battle is for the rights to the true name of who is the house of Israel. According to CI adherents, the Jews have falsely mislead the general public into believing that they are:

> The chosen people of God . . . the Jews have done this by interpreting the Bible wrong, as well as telling the big Jewish Lie, that six million people died in WWII. Six million didn't die. That whole thing is just a way for the Jews to get sympathy and get the thing they want, which in the end is world domination. We (meaning white Adamic race) are in a battle that we are going to lose if we don't wake up people real soon. The Jews control our schools, media, police, laws, you name it. This isn't a battle that the Jews are going to win. Shortly, the white Adamic race will wake up to the real agenda of the Jews and help us overthrow them.

The apocalyptic battle that CI adherents believe in originated in the Bible in the passage of Jacob and Esau. Esau and Jacob were twins, with Esau being the firstborn of the two. Esau—an "Adamite, Semite, and a Hebrew," as well as the firstborn—felt that he was heir to the birthright of his lineage (Weisman, 1991 p. 5). Normally, this would be the case, but according to Charles Weisman (1991 p. 5), the author of *Who Is Esau-Edom?*, God said that:

> Esau married daughters of Canaan or Canaanites (Gen. 28:6; 36:2), the daughters of Heth or Hittites (Gen. 26:34; 27:46; 36:2), the daughters of Ishmael (Gen. 28:9; 36:3), the daughters of the Hivites (Gen. 36:2), and had intermarried with the Horites (Gen. 36:19-21; Jasher 30:28) . . . the descendants of Esau from these marriages became known as "Edomites" or "Edom (Gen. 36:1, 9).

Because Esau "race-mixed" with what CI adherents believe are today's Jews, he fell into displeasure with his parents, as well as with God, and this is why God favored Jacob over Esau. A hostile relationship developed between the two brothers. The "antagonistic relationship between Esau and Jacob was magnified and intensified when Esau "sold his birthright to Jacob" for a mere bowl of pottage (Gen. 25:33). According to Bertrand Comparet, a well-known CI preacher in the 1950s and 1960s, the story of Esau's selling his birthright is more symbolic than most preachers want you to believe:

> What the Bible tells you about Esau selling his birthright for a bowl of stew: that isn't when he lost it. That was merely a formal ceremony by which he gave up any claim to it; but he lost it when he did the thing that rendered it impossible for him to continue as the head of the clan. His descendants from then on would be mongrelized, half satanic. So recognizing that he was already out of the line for leadership, he sold it for a bowl of stew (Comparet, n.d., *The Cain-Satanic Seedline*, p. 20).

Jacob then had the legal right to receive the blessings of the inheritance that he had received from Isaac by pretending he was Esau. When Esau realized what had happened and that Jacob now possessed the blessings that could have been his, "Esau hated Jacob because of the blessing" (Weisman, 1991, p. 58).

Like earlier interpretations, a biological approach is applied to the Bible. Like early positivist theorists of criminology, CI believers state that the bad traits that a Jew inherits can be passed on to other Jews. Wesley Swift, a leading CI proponent who wrote several books, pamphlets, and tracts, states in a sermon:

> Laws of heredity are well established. Christ recognized the laws of heredity even in the patterns of thought and conduct and he turned to that were his adversaries. I expect you to be this way. You are going to be just like your father. He was a liar, a devil, and a murderer and his offspring will be like him. So, you have a perpetual Juke family among certain people (Swift, n.d., Basic Identity).

What Swift is referring to is the study done in 1877 by Richard Dugdale, who looked to "infer that criminal (and anti-social) behavior is inherited. Dugdale was able to trace a number of criminals, prostitutes, and paupers in the family line, all derived from the original criminal father" (Dugdale, as cited in Williams & McShane, 1999, p. 37).

Charles Weisman points out that "this contrasting and conflicting ways and thinking between Jews and the white European is not artificial or a result of their environment, but is based on the nature of their physical conditions. The conflicting differences that exist between Jews and those of the white race are primarily a result of their genetic differences, or their 'nature and constitution.' Jews act and think differently from white people because there is a difference in their brains" (Weisman, 1991, p. 55–56). CI adherents believe that all of today's Jews still have this jealousy and are deliberately passing a Jewish agenda, to take over the world and "kill Israel (White Christians) for it is the only way to get the revenge for Esau" (Weisman, 1991, p. 98). According to CI adherents, the inheritance was to be the domination, or ruler, of the earth.

According to Charles Weisman, "Jacob-Israel birthright and blessings include a status of dominion in the earth with God as their head. This dominion conflicts with the Edomite Jew's plan of one-world dominion" (Weisman, 1991, p. 98). In order to get the Edomite (i.e., Jewish) plan enacted, they must destroy all the white Israelites on earth in order to negate Jacob-Israel's special status, therefore inheriting the title of being God's chosen people. The fight is seen as that of good (Aryan) and evil (Jewish). The battle is over the genetic lineage of the true house of Israel. That true genetic lineage is the true inheritor of the earth.

CI believers do not see the Jews as having a "special covenant" with the Lord. They see the battle as that of Jews (evil seed) trying to taint the Aryan (good seed) line, and they believe that the Jews confuse the masses into believing that the Jews are God's chosen people. They will do this through taking over the major institutions in the world (schools, media, government, financial, etc.).

According to CI adherents, evidence of this struggle has been seen throughout history and is shown in the writings of the Bible and the Talmud. According to CI believers, the two religions are diametrically opposed to each other, and therefore cannot exist together because Judaism "rejects the commandments of God" (Weisman, 1991, p. 54). This leads to the apocalyptic outlook that the war is between the Jews and the white race. Richard Butler states that "it is a contest of who is going to win" (Butler, 2001).

HOUSE OF ISRAEL

Israel, to CI believers, is not a land but a people. For Aryan Nations members, nationality defines who is Israel, the chosen people of God. For CI proponents, this nationality is the:

> Anglo-Saxon, Germanic, and Scandinavian Nations. . . . God first made His great promises to Abraham and repeated them to Abraham's son Isaac, and Grandson; Jacob (whose name God changed to Israel, "A Prince, Ruling With God"). Israel had 12 sons. The descendants of each son became a tribe, so that all the descendants of Dan became the Tribe of Dan, the descendants of Judah, the tribe of Judah and so on. After their long captivity in Egypt, they became one nation of 12 Tribes (like the one United States of 50 States), which continued until Solomon's death.
>
> Then the 10 Northern Tribes revolted and set up their own kingdom, keeping the name Israel, while the old, southern, two-tribe nation was called Judah. Thereafter, their histories are recorded separately, in the Books of Kings and of Chronicles, which, like the prophets, carefully distinguish between them. About 715 B.C., Israel was captured by Assyria and deported to the lands around the south end of the Caspian Sea and they never returned to Palestine"(Comparet, n.d., *Israel's Fingerprints*).

According to CI believers, the house of Israel is the recipient of all the promises from the Bible. To them, Israel is the Scandavian, Germanic people. How adherents know that Israel is the white race is seen in the promises that are made in the Bible. These promises are the following:

> THE ISRAELITES. THEREFORE, WE MUST LOOK FOR ISRAELITES AMONG THE CHRISTIANS. THE PROPHECIES AND PROMISES TO ISRAEL HAVE BEEN FULFILLED TO THE ANGLO-SAXON AND SCANDINAVIAN NATIONS!
>
> FIRST, they are Christians and have been from early times.
>
> SECOND, THEY ARE A GREAT NATION AND A COMPANY OF NATIONS: The United States is the largest civilized nation in the world; the British Commonwealth of Nations is legally "a company of nations." The Scandinavian nations, all of the same blood, can be identified by their history and their heraldry as the Tribes of Dan, Benjamin and Issachar.

THIRD, THEY ARE VERY NUMEROUS: In the last two centuries, the United States increased from a mere handful to about 200,000,000 people; in the last 3 centuries, the British Empire increased from 5,000,000 to over 70,000,000 Anglo-Saxons.

FOURTH, THEY EXPANDED IN COLONIES IN ALL DIRECTIONS, as God prophesied in Genesis 28:14, Deut. 32:8, Isaiah 54:2–3, etc. Their lands are on every continent and in every sea. No other nations had such colonies.

FIFTH, THEY "POSSESS THE DESOLATE HERITAGES" OF THE EARTH, AS GOD PROPHESIED. In Isaiah 49:8, "thus saith the Lord: in an acceptable time have I heard thee and give thee for a covenant of the people, to establish the earth, to cause to inherit the desolate heritages." Who else has so successfully developed the waste places, which were desolate when they first occupied them? Compare our own Southwestern States with what any other nation has done with similar deserts; compare the British colonies with those of nations of other races.

SIXTH, THEY ARE A SEAGOING PEOPLE: God said of Israel, "His seed shall be in many waters." Numbers 24:7 and "I will set his hand also in the sea and his right hand in the rivers" Psalm 89:25. The two greatest navies belong to the United States and Great Britain; the three greatest Merchant Marines to these two and Norway. [Note: This was written before the United States destroyed their Merchant Marine.]

SEVENTH, THEY "POSSESS THE GATES OF THEIR ENEMIES." Clearly, Genesis 22:17, refers to the "gateways" of hostile nations, the great waterways of the world. The two great Anglo-Saxon nations alone have power to close every important "gate" in the world and have done it in two World Wars.

EIGHTH, THEY MAINTAINED THE CONTINUITY OF THE THRONE OF DAVID: It has been proved that all the Kings of England, Ireland and Scotland are descendants of King David of ISRAEL, fulfilling the prophecy that "David shall never lack a man to reign over the House of Israel." Time allows me to give only a very few of the many prophecies about Israel which have been fulfilled by the Anglo-Saxon-Scandinavian people and by no others. At least 100 of them have been found. When you consider that there are more than 100 recognized nations, the mathematical odds against all of these being fulfilled by just one small group of nations, all of the same blood is billions to one. Do you think this happened by mere accident (Comparet, n. d., *Israel's Fingerprints*)?

According to this philosophy, then, Christianity did not descend from Jews. According to CI believers, Jews are against Christianity and want to have it wiped off the face of the earth. Where they get this anti-Christian stance is from one of the two major books of their religion: the Talmud. Here is an explanation of this by one CI minister:

> Judea Christianity is an oxymoron. There is nothing, no way that anyone can say that we got our religion from them. That's why I tell people to go read the Talmud and tell me that we got our religion from them. Because we don't. The Talmud is the sickest excuse for a religion on the face of the earth.

According to most CI adherents:

> The Talmud use to be oral tradition. Eventually someone wrote these things down, compiled it into a book, and called it the Talmud. Ever since that time it has been used by the Jews.

According to CI believers, the Talmud is written laws, or norms, of what Jews should do to wipe Christians and Christianity off the face of the planet. Here is an example of a few of the things that the Talmud permits, according to CI believers (11th Hour Remnant):

- Raping of Christian women because Christians are seen as inferior
- Have sex with a girl under three years of age
- Blasphemy against Jesus Christ

According to CI adherents, the Talmud allows "crimes against nature," such as sodomy, rape, and oral sex. One believer went on to give further proof that the Talmud is anti-Christian, describing a blood ritual by Jews against Christians:

> They [Jews] actually take blood [from Christian children] and mix it into their bread for their feast of purum. If you go back and look at the majority of children who disappear in this country [United States], they disappear during that period of time [Passover].

This "proof" further supports their anti-Semitism toward Jews. Some CI believers do not refer to what they are doing as anti-Semitic, however, because they believe that that is a term (Semite) that is reserved for those that are the true Hebrews of the Bible, as explained by one CI believer:

> I am not an anti-Semite. I am anti-Jew. Those people that call themselves Jews do not have one ounce of Hebrew blood in them. In order to call yourself a Semite you have to have descended from the Hebrew race; and they have not.

Because of this battle going on between Jews and Christians, some followers do not recognize any truth in what Jews have to say. They believe that there is an apocalyptic battle taking place on the soil of earth today, and feel the Jews will lie to win back what should have been theirs: the covenants and promises that God outlined in the Bible to the people of Israel.

Not all enthusiasts think this way. Some feel that the Jews are saying some truth, but that they are just trying to fool the public into believing that the Jews are the chosen people:

> They have made the statement themselves that they will tell you 90% truth to get you to believe the 10% lie. I am a Jew and I own a clothing store. I have a Pendleton shirt, it is 100% wool. Now is that a lie, because I am a Jew and I said it? No. They talk about it, does that mean we should disregard it? No anyone who studies the scripture, I mean studies the scripture, I don't see how they can study the scripture and come up with anything else.

THE LOST TRIBES

As the above section demonstrates, adherents of Christian Identity believe that the Bible describes those who are to be the "Anglo-Saxon, Scandinavian, and Germanic people" (American Institute of Theology, n.d., p. 112) as the descendants of the lost 12 tribes of Israel, and deserve all the privileges that come with this distinction. Believers contend that the 12 tribes of Israel can be further divided into two kingdoms, the North and the South. After the passing away of King Solomon about 925 B.C. the Northern Kingdom was composed of 10 of the lost tribes descended from Jacob. Approximately 150 years before the first temple fell in 585 B.C., the Northern Kingdom was completely conquered by the Assyrians. As was custom at that time, the victorious army carried the newly subject people off and recolonized the area. The 10 tribes were then "lost." Leonard Zeskind (1986, p. 13), in *The Christian Identity*

Movement, gives a quote from the British Israelite magazine *Destiny* describing this:

> When the people of the Northern Kingdom went into Assyrian captivity, they did not remain there. During the subsequent dissolution of the Assyrian power through its involvement in foreign wars, the people of Israel escaped in successive independent waves, leaving the land of their captures when the opportunity came to do so. Under different names, (Scutai, SakGoths, Massageate, Khumri, Cimmerans, Goths, Ostrogoths, Visigoths, etc.) they moved westward into the wilderness, across Asia Minor, then into Europe and eventually into the Scandinavian countries and the British Isle.

The 10 Northern tribes were nomadic in nature, and were used by the Assyrians as a slave warrior labor, and set out to conquer empires. The remaining two tribes—Benjamin and Judah—became encapsulated into the tribe of Judah, and were stationary. This would change starting in 705 B.C. with the death of King Sargon II. After his passing away, rival tribes began to attack Assyria in order to gain its wealth and landholdings. The king of Judah, King Hezekiah, ran one of the armies. According to the interpretation of CI theology, they were successful (American Institute Theology, n.d., p. 113). In 701 B.C.:

> . . .the new King of Assyria, Sennacherib, set about recovering his empire; one rebellious city after another was conquered, with the hideous cruelty characteristic of Assyria; and in 701 B.C., Sennacherib's huge army invaded the kingdom of Judah . . . none of the smaller cities of Judah were able to resist (AIT, n.d., p. 113).

At this point, the Assyrian pilgrimage began to include the 10 tribes of the North, as well as most of the people of the two Southern tribes of Judah. As the tribes moved, they began to expand along both sides of what is now known as the Caspian Sea. Along this pilgrimage, according to CI texts, groups from the tribes began to settle in areas, and these people would be the ancestors of the white race. Table 3-1 shows the names of the tribes and where their descendants are today.

Table 3-1

The 12 Lost Tribes of Israel

Country	Tribe
Denmark	Dan
Norway	Naphtali
Finland	Issachar
Germany	Judah
France	Zebulun
Italy	Gad
Holland	Reuben
Spain	Simeon
Iceland	Benjamin
Great Britain	Ephraim
USA	Manasseh
Sweden	Asher

To CI believers, the importance of the "Caucasian" race's being the true lost tribe is crucial to their understanding of self. Most adherents refer to this as the unlocking of "one's true identity:" "Once you understand your true identity everything changes. Your view of the world is never the same. You have to understand that the Jews are trying to suppress who the true lost sheep are."

JESUS

In mainstream religions, Jesus is considered the king of the Jews. According to CI adherents, this is not the case. The reason Jews claim this, according to CI believers, is so that they get special privilege in society as being the true house of God. For CI's anti-Jewish belief, there must be a removal of Jesus as a Jew. There are several beliefs among CI adherents that allow this; most of their justification is outlined via passages in the Bible. One of the most used passages is John 10:26, when "Christ said to the Jews, 'But ye believe me not, because ye are not my sheep!'" After Jesus told the Jews they were not of his people (race), he described how his true Israel "sheep" would react: "My sheep hear my voice, and I know them and they follow me."

According to CI believers, Jesus was put on the earth in order to convince those who obeyed the Jewish laws that they were following the orders of Satan, and that the true house of Israel was that of the white Anglo-Saxon Caucasian race:

> God's instructions were to keep his family tree [line] pure. There was to be no race mixing! It was intended by God that his family, put on Earth in flesh bodies but born of the spirit would through purity of spirit and obedience to Divine Law, bring order and righteousness to a world laboring under the rule of Satan (the Jews). John 8:23, " I am from above ye are from beneath." John 8:41, "Ye do the deeds of your father." John 8:44, "you have the devil for your father and the lust (desires) of your father will ye do." Jesus was not speaking to people, of his own race, when he made these accusations; for Jesus is not a Jew.

Jerome Walters (2001, p. 24–25) points out in One Aryan Nation Under God that the word "Jew" has several different meanings for Christian Identity members:

1. Jew "can a be label for someone who is from the country of Judea."
2. The term Jew can designate a distinct race of people. As long as it is not applied to Jesus or his disciples.
3. Identity doctrine uses the term "Jew" to refer to a specific ungodly religion.

The first meaning implies a specific geographic area. This is important because some adherents of CI also believe that they can use geographic descriptions to prove that Jesus was not a Jew. For instance, Wesley Swift, in *Was Jesus Christ a Jew?* states "that Galileans were not Jews [*sic*]" (Swift, n.d., p. 12). Swift goes on to show biblical proof of this by citing John 6:

> Jesus Christ had twelve disciples. One of them was a Jew [sic]; that was Judas Iscariot. All the rest came out of the Galilee, out of the household of His section. He asked them a question concerning His identity, and Simon Peter said: "Master, where shall we go? Thou hast the words of eternal life. We believe and are sure that though art the Christ." That means the Embodiment of Yahweh, the Very Son, the Embodiment of the Most High.

Swift goes on to prove his point. "Jesus said: 'Have not I chosen you twelve, and one of you is a devil?' While that may be just a name you call somebody, to Christ it is a generation of Lucifer." Swift finishes up his geographic proof of Galilee's not being a locality for Jews by citing John 7:1: "After these things Jesus walked in Galilee: For he would not walk in Jewry [*sic*], because the Jews sough to kill him" (Swift, n.d., *Was Jesus Christ a Jew?*, p. 12).

The implication of race, in the second meaning, hints that there must be distinguishable characteristics among those who claim to call themselves Jews. Several interviews with prominent Aryan Nations leaders suggested this:

> I have no problem telling who is a Jew and who isn't. That is because I have been awoken to my true Identity, that I am the true Israel. Israel as we see it is not that sand pit across the ocean. Those are the devil. The true Israel is the true race of Adam.

CI believers see the Bible as a prophetic canon, and believe that prophecy is a dualistic battle between good and evil. They believe that the Jews, as a race, are born with a disposition to lie, cheat, and feed off the true house of Israel (Adamic white race). It is because of this belief that publications such as *The Turner Diaries* and *The International Jew* play such an important part in their understanding of how the Jewish race is trying to take over the world. Jews supposedly will accomplish this by taking over all institutions of society to control the mind-set of true Israel (whites). This is summed up best by the 11th Hour Remnant Messenger:

> These Jews or Jewry were and still are the evil offspring of Satan from Cain, again through Esau and through those mixed blooded house of Judah which returned from Babylon 400 years before.

Several Aryan Nations members said to the author: "How can Jesus be a Jew? Jews don't believe in Jesus and they are fighting to destroy the house of real Christianity today." Most see Judaism as the equivalent of Satanism. Equating Judaism with the devil constructs an evil seed on the planet, and that seed wishes to destroy the Christian faith. Playing on this factor throughout the Bible causes fear among adherents. As we will see in the section on violence, this fear gets tapped into in order to convince others that something needs to be done about the demonic seed that is walking the planet.

"Proving that Jesus wasn't a Jew" allows for the entrance of conspiracy, as well as the rationale for Jesus's real reason to be here. The 11th Hour Remnant Messenger sums it up into four possible reasons:

> God the father, sent his only begotten son, Jesus Christ incarnate, to die on the cross and accomplish the following missions:
>
> 1. Kingsman redeemer—To enter into the Adamic/Hebrew Race and "purchase mankind" from the power of Satan
> 2. Forgive the sin of Adamic/Hebrew man—Die a sinless death, shedding his blood and water on the cross in obedience to God, His father [Blood=remission of sin. Water = life/agape/way of salvation]
> 3. The resurrection of Jesus Christ by God His father to reveal and demonstrate the power of God and the deity of Jesus Christ
> 4. The only hope for the Adamic/Hebrew/mankind eternal salvation

The second function of this biblical interpretation is to allow a cognitive rational for their hatred. One prominent Aryan Nations leader told me: "I could never hate for the sake of hating. A lot of the Klan is like that. To me I hate the Jews because that is what Yahweh (God) wants. It says so in the Bible." This revisionist interpretation of the Bible allows for the white Adamic race to see Jesus as "the segregationist supreme, calling the white race to pursue racial purity and 'build an Ark of Safety' for the race" (Walters, 2001, p. 26–27). Claiming that a superior power believes that the white race is superior, and should rule minorities, provides a rationale to be adopted by bigots, allowing them to justify their actions. Because God and his incarnate son made this law, how can hate mongers be doing something wrong? They do not hate, because they are doing the work of God. What they are really doing is showing their love for the white race. As we will show later, this cognitive dissonance allows for the objectification of what they see as another. Once this is done, it allows for the possibility of carrying out God's plan, becoming "the battleaxe and the sword to wipe the evil Jew off the face of the earth"(Wickstrom, n.d.).

ARMAGEDDON

According to one prominent Aryan Nation leader, the word "Armageddon" is of Hebrew origin: "It comes from two Hebrew words, Har Megedon. Literally translated this means land of the gathering."

Most CI adherents believe that the gathering for Armageddon will be the United States of America:

> So were this land of the gathering? The land for the re-gathering of the house of Israel is the U.S. Some say that the Jews going back to Israel is prophecy fulfilled. The only prophecy that it's fulfilling is the fact that they are being gathered back there for the burning. They will be destroyed in the very land were the Esau-Edomites killed Yashua. God told David that he would set his descendants in a new land. One that they would not have to move from again. David was standing in Palestine when he said that in a new land it would have to be different than Palestine. He told Jerusalem is not the New Jerusalem. The New Jerusalem is in the United States.

The prediction of America as the New Jerusalem is not based on just hearsay or from a crystal ball; the justification comes from biblical prophecy:

> The bible refers to the New Jerusalem as a nation being born in a day. What nation on his planet was born in a day? This one, July 4th, 1776. It also is described as being bordered by two seas and divided by many rivers.

To some CI adherents, the battle of Armageddon is taking place as we speak. Some saw the incidents of September 11 and the constant battles between Israel (the Jewish state) and the Muslims as a sign of the time of the beginnings of an apocalyptic battle.

REFERENCES

Aryan Nations. *Aryan Nations Web site* [chapter opening quote]. Retrieved April 8, 2002, from http://www.twelvearyannations.com.

Aho, J. (1990). *The Politics of righteousness.* Seattle: University of Washington Press.

American Institute of Theology. (n.d.). *AIT bible correspondence course.* Harrison, AR: AIT.

Barkun, M. (1997). *Religion and the racist right: The origins of the Christian Identity Movement.* Chapel Hill: University of North Carolina Press.

Butler, R. (Producer). (2001). *My side of the story.* [Videotape]. Coeur d'Alene, ID.

Comparet, B. (n.d.). *The Cain-Satanic Seedline.* Retrieved September 8, 2007, from http://www.churchoftrueisrael.com/comparet/compcainsatan.html.

Comparet, B. (n.d.). *Israel's Fingerprints.* Retrieved September 8, 2007, from http://www.churchoftrueisrael.com/comparet/comp4.html.

11th Hour Remnant Messenger. (n.d.). Unpublished brochure.

Gayman, D. (n.d.). *The two seeds of genesis 3:15.* USA.

Post, J. (1984). "Notes on a psychodynamic theory of terrorist behavior." *Terrorism: An international journal* 7: 241–256.

Post, J. (1990). "Terrorist psycho-logic: Terrorist behavior as a product of psychological forces." In W. Reich (Ed.), *Origins of terrorism* (p. 25–40). Cambridge, UK: Cambridge University Press.

Ridgeway, J. (1995). *Blood in the face.* New York: Thunder Mountain Press.

Swift, W. (n.d.). *Was Jesus Christ a Jew?* Retrieved June 16, 2008, from http://www.churchoftrueisrael.com/swift/swift1.html.

Swift, W. (n.d.). *Basic identity.* [Cassette recording].

Walters, J. (2001). *One Aryan nation under God.* Naperville, IL: Sourcebooks.

Weisman, C. (1991). *Who is Esau-Edom?* Burnsville, MN: Weisman.

Wickstrom, J. (n.d.) *Children of light vs. children of darkness.* [Cassette recording].

Williams, F. P. & McShane, M. D. (1999). *Criminological theory.* [Third edition]. Englewood Cliffs, NJ: Prentice Hall.

Zeskind, L. (1986). *The Christian Identity movement.* Atlanta, GA: Center for Democratic Renewal.

CHAPTER 4

Enemy of My Enemy

Kevin Borgeson and Robin Valeri

Since September 11, 2001, most of America's concerns about terrorism have focused on countering threats from groups outside the United States. However, many of the most infamous terrorist attacks in the United States—including the Oklahoma City bombing by Timothy McVeigh, bombings of abortion clinics by Eric Robert Rudolph, and the mass murder committed by Bufford Furrow—have been committed by Americans. These individuals were or are either adherents of Christian Identity, members of the Aryan Nations, or both. A recent article in *Intelligence Report* (Blejwas, Griggs, & Potok, 2005) states that over the past decade, there have been an estimated 60 terrorist acts within the United States committed by individuals in hate groups. According to the report, most of these domestic terrorist acts have religious underpinnings (see Chapter 3).

Although this book focuses on issues related to domestic terrorism, the current chapter provides a startling example of why terrorism can no longer be viewed as either a domestic or a foreign threat. Just as improvements in communication, technology, and travel have made it easy for the average citizen to develop a global network of friends, these same resources have made it easy for terrorists to develop a global network of hate. Hate has been online since the early 1980s, providing hate groups with an inexpensive means for getting their message out and for communicating with like-minded people throughout the world. Don Black, a former Klansman and creator of Stormfront.org, the first extremist hate site on the Web, is quoted by Schneider (1995) to this effect:

> All of this [Internet] has had a pretty profound effect on a movement whose resources are limited. . . . With the phenomenal growth of the Internet, tens of millions of people have access to our message. . . . The access is anonymous and there is unlimited ability to communicate with others of a like mind.

Today, Stormfront.org serves as a global resource for white supremacists, providing links to such groups in Canada, Britain, Europe, Russia, Serbia, South Africa, and Australia, to name only a few. Stormfront.org offers advice on legal issues, domestic concerns, and dating, as well as providing resources and links for activists. Whereas some hate sites are quite blatant, others are not quite so obvious in their expressions of hate and may purposefully tone down their use of defamatory rhetoric. Consequently, people are not always able to determine that a specific Web site is a hate site, and thus may not factor this into their processing of the information presented (Borgeson & Valeri, 2004; Valeri & Borgeson, 2005).

The current chapter provides one example of a growing alliance between a domestic terrorist group and a foreign terrorist group. Specifically, this chapter focuses on the growing alignment between the Aryan Nations, one of the most notorious hate groups in American history, with one of its former enemies, followers of Islamic Jihad.[1]

RECENT CHANGES WITHIN THE ARYAN NATIONS

Anti-Jewish Sentiment

According to Kathleen Blee (2002), women who join organized racist groups learn to become anti-Semitic. Although most Aryan Nations members interviewed for this project were male, Blee's assertion is relevant because most of the interviewees started off in other organized groups hating blacks, not Jews.[2] It was their journey into hate that allowed most of them to "come to the realization that the Jews are to blame for most of the world's problems." Take, for instance, this quote

1. During the interviews, members of the Aryan Nations referred to the Muslims they had contact with as "Islamic Jihadist" because they did not want to reveal the identities or locations of these individuals or groups. For the purpose of this chapter, the authors will use this same terminology: "Islamic Jihadist."
2. The average person in the Aryan Nations has been in approximately three hate groups before joining the Aryan Nations.

from a man who joined the Aryan Nations after belonging to several other white supremacist organizations:

> I have been in several organizations over the years—skinheads, Klan, you name it. The problem is they have it all wrong. They want you to believe that the blacks and Hispanics are to blame for society's problems; they aren't. They are too dumb to figure it out. Once I realized that, I figured somebody bigger had to be behind all the problems in the U.S. It wasn't until I met [a friend] that I began to understand that the Jew was behind most of it, and they are using the blacks and Christians to get their agenda across.[3]

Part of the transformation in their target of hatred is a direct result of their belief in a Jewish conspiracy. For members of the Aryan Nations, the Jewish people serve as scapegoats and are to blame for a host of problems. One of the most frequently expressed beliefs was that Jewish people working in the World Trade Center knew in advance of the planned attack and, therefore, did not go to work on that day. One member stated that the number of people in the buildings did not add up, and the only way that could happen is if the Jews who worked in the building had told other Jews not to go to work on that specific day.

> . . . just like the number, six thousand . . . six thousand people died in the World Trade Center bombings . . . that's not true because just recently the number's down to three thousand seventy . . . three thousand seventy . . . now you take that building between what . . . quarter to nine and nine thirty or quarter after nine . . . in the morning, do it . . . do the . . . just think about how many people usually generally work? . . . worked in them buildings . . . fifty thousand a building I believe the number was . . . now nine o'clock in the morning, quarter after nine . . . nine thirty, whenever it happened to be. I don't remember now . . . figure in how many were there, approximately how many were there on any other given day, then factor in how many were there and how many were killed and how many had a . . . really had a chance to get out of the building . . . out of the area before the buildings came down . . . you know and I know you could all of that by factoring you know an average you know and the numbers aren't gonna [*sic*] jive even that six hundred . . . even that six thousand that was only like twenty five people a floor you know and I was told by people that work there . . . or had worked there in the past, you know, electricians or whatever.

---

3. Unless otherwise specified, all quoted material in this chapter stems from interviews conducted with members of the Aryan Nations, all of whom were guaranteed anonymity. The interviews were conducted between August 2001 and December 2005.

Another common view is that Jews control the media. In the Aryan Nations, "media" is a buzzword for "Jew." Most members cling to a stereotype that started with the Protocols of the Elders of Zion, which stated that Jews have a secret desire to control most social institutions, including the media, banks, schools, and government. One of the first conspiracies that one of the authors (Borgeson) was subjected to upon entrance into this area of research was that the Jewish-controlled media covered up the Jews' involvement in 9/11:

> How come immediately after 9/11 there was [*sic*] pictures floating around of Jews dancing on top of buildings? And within minutes the pictures were never seen again? That's because the mainstream Jew media wanted them hushed up so no one knew they were behind it. Are those conspiracies? No, I think they're actual fact.

The Aryan Nations as a Christian-Based Organization

One important distinction made by most interviewees between Jews and non-Jews is religion. According to the interviewees, religious differences make Jews intrinsically different. The Aryan Nations is based on Christianity. Racism and anti-Semitism, for most Aryan Nations members, have been justified through Christian Identity, the cornerstone of the Aryan Nations movement for over 25 years. Members of Christian Identity believe that the real chosen people—the 12 lost tribes—are the white Anglo-Saxon races of the world. Christian Identity adherents believe that there is currently a battle between good and evil taking place on this earth, and that all whites must unite against the Jew to get their chosen rights back when Jesus returns. For most members, race and anti-Semitism are so tied in with Christian Identity that to step back from the religion is seen as becoming soft on race and anti-Semitism, which the founder of the organization, Richard Butler, believed in:

> I believe that the Identity message is the only religion that should be allowed. You can't let a bunch of sand niggers into the organization—you are supposed to be about the Aryan race. Last time I looked, they sure the hell were not Aryan and they shouldn't be allowed to even be in this country.

The above quote reflects two important beliefs. The first is that Christian Identity is the only religion. Consequently, people with any other religious ideology should not be allowed to join the Aryan Nations. The second is that all Muslims are black, and therefore should not be allowed to join. When an interviewee was informed that not all Muslims are dark skinned, he replied:

> I haven't seen any. And even if there is some that are white, they should be shot for disserting [*sic*] their race. Whites are the chosen descendants of the bible, and whites need to wake up and reclaim back our heritage from the children of Satan.

As the above quote demonstrates, race is inherently tied to Christian Identity, and the movement believes that all people of the white race should be Christian. Thus, whites who do not adhere to a Christian faith are traitors to their race. Christian Identity is so tied into the belief system that some adherents see the addition of any other religion, in this case Muslims, as part of the Aryan Nations mission as a sin, which should not be tolerated:

> Because if they were to interview me, like you are now, they would found out that we do not let anyone in the organization that is not Christian. Period. There are a lot of groups out there that claim to be Identity, but they will let anyone join as long as they are open to the message. That is not good enough because all down through scripture every time we have aligned ourselves with heathens, pagans, or unbelievers God has taken his blessing off of us and we have failed.

Changing Religious Views

For some members of the Aryan Nations, it is not the specific beliefs of Christian Identity that are important to the cause, but the religious justification. One member expressed this by stating: "I could never hate someone for the sake of hating them. I base my views on the Bible. The typical race-hating thing gets you nowhere." As a result of this change in position on diversity of religion, the Aryan Nations leadership has decided to allow Muslims into their organization. As a matter of fact, they say that by allowing different religions, they are living up to the "mission statement" of founder Richard Butler:

> As was the stance of the founder, Pastor Richard G. Butler, Aryan Nationss [*sic*] was developed as an institution of Aryan Virtue, along with the Aryan Brotherhood, which coalesced as a splinter within the

> prison systems, becoming the foremost means of protection for Aryans whose freedom has been confiscated. Pastor Butler's conception was never an Identity-only organization, and it is not today, however, as people in the past visited the "Church of Jesus Christ - Christian" in Hayden, Idaho, it was all too clear that the group's nucleus was the Identity Belief, as it still is today. ... The decision was made to make the Aryan Nationss [*sic*] two distinct entities, the first being the Aryan Nationss [*sic*] Proper, and the second, the Tabernacles of the Phinehas Priesthood, the Biblical wing of this Organization, this was done because many Identity adherents sought their own group to themselves, while our views and systems were attracting the most virulent and therefore the most worthwhile membership, although they were not necessarily Biblical Believers. Butler would not turn these men away, and neither will we (Aryan Jihad).

In response to the flood of e-mails the organization received, those in charge of the Aryan Nations Web site posted a clarification notice describing their nonreligious stance, as well as their belief that the Islamic Jihadeen are biblically supported:

> There has been a little misunderstanding as to what our perspective is as far as an alliance with Islamic Jihadeen, and our own Phinehas Priests. There are some out there who would like to imply that we are now an Islamic Fundamentalist Organization, and this is erroneous, our Organization is not for the support of any religion, however based on the history of the Aryan Nationss [*sic*], the rules and conduct are based on Biblical Law, and the general views of the bulk of this organization is the acceptance of Aryan Messianic Identity, or other forms of what is called "Christian Identity" in most circles (Aryan Jihad).

Responses to the above two postings reflected such great interest in the Aryan Nations by non-Christian Identity members that the leaders of the Aryan Nations are planning to make liaison appointments of different religions within the organization.

Shared Hatred

When asked why the Aryan Nations was aligning itself with Islamic Jihadists, interviewees were quick to point out that the current Christian Identity movement was dead, and that the movement lacked spine:

> The white Christians in today's movement don't have the fervor that the Muslims have. They wouldn't lay down their life for a cause. As a matter

> of fact, throughout most of the movement, there have been very few people who would lay down their life and be a martyr. If we could instill that pride back into our race the Jews wouldn't have a chance against us. We would be fighting with such veracity they wouldn't be able to stop us.

In response to the dying Christian Identity movement in America, Aryan Nations members began to develop a camaraderie with those who they felt had more "spine"—namely, Islamic Jihadists:

> There is a strong brotherhood with Muslims. Doesn't matter where you live, you are going to go to the aid of another Muslim. It is the way that Christianity use to be before they brought it to the whole world. That was our religion, the white races, for his children on the earth.

As the Aryan Nations began to branch out in its religious denominations, members began to see that they shared similar group characteristics: analogous moral values, perceived media persecution, and hatred for the United States government. The following quote reflects the belief that adherents of Islam have moral values similar to those of the Aryan Nations:

> Further, seeing the errors of the past, we have taken this approach with alliances to Islamic adherents, because we find their standards of morality to be nearly analogous to our own, and their resolve to uproot and destroy the fallen tree of the Garden, the satanic "jew," to also be analogous to our own desires and devotion. In this sense, Islam is our ally. . . . Islam has not been dishonored as much by "jew"ish incursion, therefore Islamic Jihadeen have safeguarded the purity of the very instinct for self preservation for which we hold the most vociferous esteem (Aryan Jihad).

The quote then continues with a statement that reflects the belief that the Jews control the media, and as such are able to present a negative view of Muslims:

> The "jews" have used their control of the media to portray the Islamist as a wild animal, the subhuman of the world, as they preach the virtues of themselves along with all races and false creeds that are ignoble enough to collaborate with them. Whereas the Aryan realizes that Ishmael was blessed by Yahweh, and prophesied to always be the fist in the face of his enemy. Ishmael's descendants are among the Arabs and Israelites today, and far be it from us to say that Yahweh was wrong, and the "jews" are truthful (Aryan Jihad).

The next part of the message explains why followers of Islam are the allies of the Aryan Nations:

> The Ishmaelite is our ally because he is the fist that takes the first strike against Satan's spawn, the "jew." The Aryan is the knock out punch, the Children of Yahweh will destroy His enemies, and Ishmael will stand aside his brother Isaac and receive his share of the glory and honor. What Yahweh has blessed, is blessed as far as Ishmael, and what Yahweh has cursed and condemned to death, as far as the demonic seed falsely known as the "jew," all of the spiritual sodomites of this world along with their minions of deceived zionist [*sic*] collaborators, all of their horses and all of their men will not put humpty dumpty [*sic*] "the jew" back together again (Aryan Jihad).

Then the message goes on to reassure those members of the Aryan Nations who believe that Christian Identity is at the center of the organization of its importance, while asserting that members of other religions will be welcomed into the Aryan Nations:

> The Tabernacles of the Phinehas Priesthood will remain the Biblically based foundation of the Aryan Nations, however, membership of the Aryan Nations will continue to welcome Islamists and adherents to other moral religions into our ranks. There is nothing new being done within the Aryan Nations; all of our brothers and sisters were always welcome to join our ranks, and as it was in the past, the Biblical views of the Aryan Nations conception will remain the basis of our structure (Aryan Jihad).

Lastly, the message explains that the two groups view jihad somewhat differently, but both still show superiority to those of the Jewish faith:

> There is one differentiation that the Islamic Believer must accept that is not necessarily Islamic, and this is our modus operandi regarding Jihad. Islamic Jihad is the cleansing of Islamic controlled lands, once the land is decontaminated; the Jihad is over, on the other hand, our Jihad is more specific, and it is strictly and primarily for the extermination of the satanic "jew," worldwide. We see no borders, nor boundaries. Any country found to be allowing the "jew" to exist, will be at the receiving end of this Jihad, they will find no sanctuary, as the freedom fighters of Islam have been denied sanctuary even from their own "brothers." Whoever harbors or aides the "jew" will be considered a "jew," thus receiving the same fate, or worse. Aryan Nations respects no borders, nor any political system that will collaborate or hide our enemies, the genocide of satanic "jewry" is the foremost thrust of our Jihad, and this is the main purpose of the Aryan Nations.

Both groups also share a vile hatred for the U.S. government. Their loathing is based on the conspiratorial belief that the "Jewish-run United States government" is trying to control citizens of both countries:

> The U.S. is trying to destroy their [Islam] religion. They [United States] have twisted their words to suit their cause [world domination for Jews]. The Jewish-run government is trying to control Islam, like they did to us. For those we talk to, they want to go back to "true Islam." You see they are like those in Identity—we say "true Christianity." We both have a lot in common, and we shouldn't fight with each other. We both are forms of religions; biblically we should not be fighting because we both recognize one god.

By drawing parallels between how the United States has treated their own group and how the United States is handling Islam, the Aryan Nations leadership believes that it can help Jihadists—because the Aryan Nations has already "gone through what they are currently going through."

Hate, Morality, and Christian Identity

An examination of Christian Identity from the perspective of the *Duplex Theory of Hate* (Sternberg, 2005) reveals that Christian Identity serves an important function for the political activity of the Aryan Nations by providing the rationale for hate and/or prejudice and a means of socializing people in these beliefs.

According to Sternberg's (2003; 2005) *Duplex Theory of Hate*, hate is neither the opposite of love nor the absence of love, but is related in a psychologically complex manner to love. Similar to love, there are three components to hate: intimacy, passion, and commitment. However, unlike love, the basis for hate is the negation of each of these three components. Just as there are love stories, there are hate stories, and these stories explain why the hated individual or group should be avoided, feared, and viewed with contempt (Sternberg, 2005). Sternberg suggests that hate stories achieve this through five steps: (1) by revealing the target as anathema; (2) by uncovering the target's plot or plans against the in-group; (3) by exposing the extent of the target's power; (4) by demonstrating that the target has put its plan into action; and (5) by revealing the extent to which the target has succeeded in achieving its goals. Christian Identity accomplishes this by portraying Jews as less than human and as deceitful, and therefore deserving of contempt.

The Serpent Seed tenet of Christian Identity, by identifying Jews as the offspring of the devil, provides a clear rationale as to why the Jews are worthy of hatred. According to Christian Identity, the Jews are not God's chosen people; instead, the true inheritors, as stated in the Bible, are "the White Anglo-Saxon Germanic people of the world." Christian Identity adherents believe the Jews are trying to deceive the world into thinking that the Jewish people are the true inheritors of God's promises, rather than the Anglo-Saxon race. Thus, the plot by the Jews to deceive the world is revealed. According to Christian Identity, over the centuries, the Jews have not only misinterpreted the Bible, but have also successfully deceived millions into believing that the Jews are the chosen people. To members of the Aryan Nations, this is clear evidence that the Jews are a powerful group, have translated their plot into action, and are achieving success in accomplishing their goals. Finally, and as demonstrated in the preceding quotes, members of the Aryan Nations believe that the United States, because of the current tide of political correctness combined with the liberal teachings in schools, provides a fertile ground for the Jews to spread their agenda of world domination. Thus, Christian Identity preaches the core elements of hate. It not only teaches its followers that the Jews are worthy of hatred and fear because they are the offspring of the devil, but also teaches that Jews should be hated and feared because they have been successful at deceiving millions and, given the current political climate, will continue to be successful in gaining world domination.

Another way to understand the role that Christian Identity plays in justifying hatred and prejudice toward Jews is that the very beliefs of CI make hatred and prejudice toward Jews morally acceptable. Bandura (2005) states: "Moral justification sanctifies violent means." According to Bandura (2005), people who have been brought up to view killing and violence as morally reprehensible do not suddenly come to believe that killing and violence are good, but instead redefine the morality of killing so that violence toward a particular group becomes justified. As Bandura points out, religious beliefs are often used to justify such behavior—whether it is Pope Urban II justifying the Crusades as doing the Lord's work, or Islamic extremists proclaiming terrorism as a necessary means for fighting the evil spread by decadent infidels. Christian Identity, by proclaiming Jews as the offspring of the devil, morally justifies hatred and violence toward them.

Similarly, Opotow (2005) posits that individuals or groups who are viewed as evil enemies are seen as falling outside the scope of justice or undeserving of the justice and fairness extended to one's own group. For moral exclusion to occur, Opotow (2005) explains that hate must be directed toward a group rather than toward an individual, and must be nurtured and encouraged by the culture or ideology of the group. Finally, a cause-and-effect link must be established that justifies violence toward the excluded group. Once a group is morally excluded, there is no longer any obligation to help or protect the excluded group. Thus, actions that would be deemed reprehensible if directed toward one's own group are condoned when directed toward the excluded group. In this way, moral exclusion not only justifies hate toward a specific group, but also provides a rationale for violence toward the group. As described in this and in previous chapters, Christian Identity specifically targets Jews as the enemy, encourages hatred toward this group, and establishes several reasons for striking out against this group. Because the Aryan Nations was built around the beliefs of Christian Identity, it has provided the rationale for violence toward Jews or Jewish-controlled institutions.

Group Position as the Basis for Prejudice Toward Jews

According to Blumer (1958), a "sense of group position" is the basis for prejudice. Blumer suggests that prejudice can best be understood from a relational perspective, where one's own group should stand in the social order in relation to another group. According to Blumer (1958), the dominant group believes it is superior to the subordinate group; believes the subordinate group is intrinsically different from them; believes its own group, the dominant group, has proprietary claim over certain rights or resources; and believes the subordinate group desires a greater share of these rights and resources. Blumer (1958) further explains that the feelings of superiority over and distinctiveness from the subordinate group provide the basis for the dominant group's aversion toward the subordinate group, and that the feeling of proprietary claim to rights and resources, coupled with the belief that the subordinate group is encroaching on these resources, together create the dynamic force behind prejudice.

If Blumer's framework of prejudice is applied to the beliefs of the Aryan Nations, it is clear to see that members of the Aryan Nations see themselves as both superior to and intrinsically different from Jews. This is especially evident in the core belief of Two Seed Line Identity, in which Aryans are seen as descendants of Adam, whereas Jewish people are viewed as the descendants of the devil. Additionally, based on the tenets of Christian Identity, Aryan Nations members have proprietary claim not only to the United States, but also to the world, because they view themselves as the rightful inheritors of the planet. However, interviewees consistently expressed the fear that the Jews were slowly encroaching on their resources—specifically, that Jews were taking over the United States, its government, media, and financial institutions, and were brainwashing the public. As stated by Blumer, it is this final piece—that the subordinate group, the Jews, is trying to take what rightfully belongs to the dominant group, the Aryan race—that creates the dynamic force for prejudice.

Shared Goals and Reducing Prejudice Toward Muslims

Gordon Allport in *The Nature of Prejudice* (1954) provides recommendations for the reduction of prejudice. These recommendations, because they share the idea of bringing together members of different groups, have come to be known as the contact hypothesis. In Allport's view, for intergroup prejudice to be reduced through contact, certain situational conditions are necessary. Specifically, prejudice is likely to be reduced when contact between the groups is prolonged and focuses on achieving a shared goal, when status between the groups is equal, and when the new policy of integration is officially or institutionally supported.

An example of reducing prejudice through these means is provided by a classic study conducted by Sherif and colleagues (Sherif, Harvey, White, Hood, & Sherif, 1961). In this study, stereotyping and prejudice were first created between the Eagles and Rattlers, groups of boys at summer camp, and then were later reduced by creating situations in which the two groups needed to work together to solve a problem, repairing the camp's water system and fixing a broken-down truck.

Whereas the findings of Sherif and colleagues suggest that given the right conditions, prejudice can be reduced through contact, it is also possible that efforts to unite the Aryan Nations and followers of Islamic Jihad could result in factionalism or horizontal hostility (Blake, Shepard & Mouton, 1964; Brewer, 1999; Brewer, 2000; Brown & Wade, 1987; Deschamps & Brown, 1983; White & Langer, 1999). Research by White & Langer (1999) suggests that horizontal hostility can arise when members of a minority group believe that the addition of new members will devalue the in-group identity by making it less distinctive. Similarly, research by Brown and colleagues (Brown & Wade, 1987; Deschamps & Brown, 1983) suggests that people do not always react positively when boundaries between their groups are diminished. The results of research by Brewer (1999, 2000) also suggest that to the extent to which in-group boundaries and intergroup distinctions engender feelings of identity and loyalty to one's group, cooperation with an out-group can threaten an individual's social identification. Thus, contrary to the contact hypothesis, members of the Aryan Nations may resist attempts by Aryan Nations leaders to form an alliance with followers of Islamic Jihad.

CONCLUDING COMMENTS

The present chapter examined the prejudice of Aryan Nations members toward Jews. Consistent with theories of hate, moral exclusion, and prejudice, Aryan Nations members not only view themselves as inherently different from Jews, but also as superior to them. Specifically, Aryan Nations members believe that religious differences make Jews not only different from them, but also inferior to them. According to Christian Identity, white Anglo-Saxons, not the Jewish people, are God's chosen people. The tenet of the Serpent Seed further suggest that Jewish people are Satan's offspring. Additionally, it is evident that members of the Aryan Nations see themselves as having certain proprietary claims to the United States or to things that are American, and believe that Jews are not only encroaching on, but usurping these rights and resources. For example, members of the Aryan Nations believe that the government of the United States as well as the media are controlled by Jews, and that Jews are using the media to shape and control America's views, presenting Jews in a positive light and Muslims in a negative light. Thus, Aryan Nations members' attitudes and orientation toward Jews can be understood from the perspectives of hate, moral exclusion, and prejudice.

The present chapter also examined the growing alliance between the Aryan Nations and followers of Islamic Jihad. According to the contact hypothesis, for prejudice to be reduced between two groups, one necessary component is a superordinate goal. The Aryan Nations and Islamic Jihadists share two common enemies and therefore two common goals.

As stated previously, the media or perceived media persecution provides one common threat to both groups. Members of the Aryan Nations and Islamic Jihadists each believe that the media presents their group in a biased and unfair light. The second, and stronger, shared threat is the Jew. Defeating the Jew provides the two groups with a strong superordinate goal and basis for an alliance.

However, as is evident from the contrasting viewpoints expressed by interviewees as to whether the Aryan Nations should remain a solely Christian organization, there is a growing difference among members of the Aryan Nations with regard to an alliance with Islamic Jihadists and the role of Christian Identity in their organization. Some members of the Aryan Nations believe that their organization and its members should remain tied solely to Christian Identity. Others believe that people from other faiths—specifically Muslims, who share the Aryan Nations' goal of defeating Jews—should be allowed to join the Aryan Nations. Leaders in the Aryan Nations have tried to satisfy people on both sides of the argument by dividing the Aryan Nations into two distinct entities: a political entity, Aryan Nations proper, and a religious identity, the Tabernacles of the Phinehas Priesthood, and also by expressing respect for the followers of Islamic Jihad.

Given that not all members of the Aryan Nations are in agreement about the centrality of Christian Identity to the organization or about allowing Muslims to join the Aryan Nations, it is predicted that consistent with previous research (Blake, Shepard, & Mouton, 1964; Brewer, 1999; Brewer, 2000; Brown & Wade, 1987; Deschamps & Brown, 1983; White & Langer, 1999), infighting within the Aryan Nations will increase, leading to factionalism or horizontal hostility. As mentioned previously, group members do not always respond positively when boundaries between groups are diminished, especially if the new members would devalue the in-group identity (White & Langer, 1999). Consequently, group members may resist working toward a superordinate goal in an attempt to preserve their in-group distinctiveness.

The contact hypothesis suggests that working toward a superordinate goal will serve to reduce prejudice between two groups only if they are successful at achieving that goal (Blanchard, Weigel, & Cook, 1975; Worchel, Andreoli, & Folger, 1977; Worchel & Norvell, 1980). The results of research by Worchel and colleagues (Worchel, Andreoli, & Folger, 1977; Worchel & Norvell, 1980) suggest that when cooperation between two previously competitive groups is unsuccessful, the out-group will be blamed for the failure unless there is an obvious and alternative cause for the failure. Given that it is highly unlikely that the Aryan Nations and Islamic Jihadists will be successful at accomplishing their shared goal of defeating the Jewish people, the alliance between these two groups is likely to be unstable and short lived. As Brewer (1999, 2000) points out, two groups working together to achieve a shared goal require trust. In the face of failure, especially when there is contempt or fear of the out-group, blaming and scapegoating are likely to occur.

Lastly, the present chapter examined how some members of the Aryan Nations are able to reconcile their definition of race with their recent alliance with Islamic Jihadists. The interviews and ethnographic data support the hypothesis that participation with Islamic Jihadists does not change the racial views of Aryan Nations members. The available literature on white supremacists has a tendency to focus on "collective social-psychological factors" (Blee, 1996), suggesting that individuals keep the same rigid beliefs throughout a lifetime (cf. Adorno, Frenkel-Brunswik, Levinson, & Sanford, 1950). Although these data are consistent with that supposition, they also show the complexity of beliefs that members share, and how exception to the racist rules can be made. The field of cultural studies (cf. Grossberg, Nelson, & Treichler, 1991; Turner, 2002) extends the first proposition. Cultural studies goes further than the social-psychological exploration of discrimination and prejudice by suggesting that the culture of the Aryan Nations and the orientation and behavior of its members are less overtly structured, and consequently allow for changes in ideology to fit "their mission." Behavior among individuals is not homogenous, and such ideological differences cause friction among the members. The present research suggests that the Aryan Nations ideology is more fluid than previously thought, and is changing. Part of the change is due to infighting within current white supremacist movements, juxtaposed with lack of action by "white kindred." The recent alliance with Islamic Jihadists has led to the development of a new social movement ideology that allows the white supremacist movement to remain alive in the United States. Whereas an alliance with Islamic Jihadists is consistent with the anti-Semitic views of the Aryan Nations, practices of selective adoption (Blee, 1996) are necessary for Aryan Nations members to remain consistent with their racist views.

In conclusion, the prejudicial views of the Aryan Nations toward Jews are consistent with various models of hate, moral exclusion, and prejudice. Although there is a growing alliance between certain members of the Aryan Nations and Islamic Jihadists, it is predicted that this alliance will continue to cause infighting within the Aryan Nations, and may lead to the creation of splinter groups within the movement. Thus, it is predicted that an alliance between the Aryan Nations and Islamic Jihadists will be an unstable one.

REFERENCES

Adorno, T. W., Frenkel-Brunswik, E., Levinson, D. J., & Sanford, R. N. (1950). *The authoritarian personality*. New York: Harper & Row.

Allport, G. W. (1954). *The nature of prejudice*. Cambridge, MA: Addison-Wesley.

Aryan Jihad. Retrieved October 18, 2005, from http://aryannations.org/aryan_jihad.htm.

Bandura, A. (2005). "The role of selective moral disengagement in terrorism and counterterrorism." In F. M. Moghaddam & A. J. Marsella (Eds.), *Understanding terrorism: Psychosocial roots, consequences, and interventions* (p. 121–150). Washington, DC: American Psychological Association.

Blake, R. R., Shepard, H. A., & Mouton, J. S. (1964). *Managing intergroup conflict in industry*. Houston, TX: Gulf.

Blanchard, F. A., Weigel, R. H., & Cook, S. W. (1975). "The effect of relative competence of group members upon interpersonal attraction in cooperating interracial groups." *Journal of personality and social psychology* 32: 519–530.

Blee, K. (1996). "Becoming a racist." *Gender and society* 10 (December): 680–703.

Blee, K. (2002). *Inside organized racism: Women in the hate movement.* Berkeley: University of California Press.

Blejwas, A., Griggs, A., & Potok, M. (2005). "Terror from the right." *Intelligence report* 118(Summer): 33–46.

Blumer, H. (1958). "Race prejudice as a sense of group position." *Pacific sociological review* 1: 3–7.

Borgeson, K. & Valeri, R. M. (2004). "Faces of hate." *Journal of applied sociology* 21: 99–111.

Brewer, M. B. (1999). "The psychology of prejudice: Ingroup love or outgroup hate?" *Journal of Social Issues* 55: 429–444.

Brewer, M. B. (2000). "Superordinate goals versus superordinate identity as bases of intergroup cooperation." In D. Capozza & R. Brown (Eds.), *Social identity processes: Trends in theory and research* (p. 117–132). London: Sage.

Brown, R. J. & Wade, G. S. (1987). "Superordinate goals and intergroup behavior: The effects of role ambiguity and status on intergroup attitudes and task performance." *European journal of social psychology* 17: 131–142.

Deschamps, J. C. & Brown, R. J. (1983). "Superordinate goals and intergroup conflict." *British journal of social psychology* 22: 189–195.

Grossberg, L., Nelson, C., & Treichler, P. (1991). *Cultural studies.* New York: Routledge.

Opotow, S. (2005). "Hate, conflict, and moral exclusion." In R. J. Sternberg (Ed.), *The psychology of hate* (p. 121–153). Washington, DC: American Psychological Association.

Schneider, K. (1995, March 13). "Hate groups use tools of the electronic trade." *New York Times* p. A12.

Sherif, M., Harvey, O. J., White, B. J., Hood, W. R., & Sherif, C. W. (1961). *The robbers cave experiment: Intergroup conflict and cooperation.* Middletown, CT: Wesleyan University Press.

Sternberg, R. (2003). "A duplex theory of hate: Development and application to terrorism, massacres, and genocide." *Review of general psychology* 7: 299–328.

Sternberg, R. (2005). "Understanding and combating hate." In R. J. Sternberg (Ed.), *The psychology of hate* (p. 37–50). Washington, DC: American Psychological Association.

Turner, G. (2002). *British cultural studies: An introduction.* New York: Routledge.

Valeri, R. M. & Borgeson, K. (2005). "Identifying the face of hate." *Journal of applied sociology* 22: 91–104.

White, J. B. & Langer, E. J. (1999). "Horizontal hostility: Relations between similar minority groups." *Journal of social issues* 55: 537–560.

Worchel, S., Andreoli, V. A., & Folger, R. (1977). "Intergroup cooperation and intergroup attraction: The effect of previous interaction and outcome of combined effort." *Journal of experimental social psychology* 13: 131–140.

Worchel, S. & Norvell, N. (1980). "Effect of perceived environmental conditions during cooperation on intergroup attraction." *Journal of personality and social psychology* 38: 764–772.

CHAPTER 5

Counterterrorism

Stephen E. Costanza, John C. Kilburn Jr., and Ronald Helms

A BRIEF HISTORY OF COUNTERTERRORISM

Any effective history of counterterrorism suffers from the burden of having to devise a description of a subfield of law enforcement that has never been adequately defined. Even after reviewing academic and popular literature, it is hard to pinpoint one accurate definition of terrorism. The Federal Emergency Management Agency (FEMA) defines terrorism as the ". . . use of force or violence against persons or property in violation of the criminal laws of the United States for purposes of intimidation, coercion, or ransom." However, because this definition is broad, and claims coverage of both domestic and international terrorism, it gives room for speculation about moral connotations of actions. One party's action may be seen as evil, whereas other actions may be seen as heroic depending on the time, place, and circumstances. For example, using the FEMA definition, one can argue that several integral world events throughout history—such as the Boston Tea Party, the storming of the Bastille, Nat Turner's Rebellion, the Munich Beer Hall Putsch, the Confederate invasion of Fort Sumter, or even the Tet invasion in Vietnam—could all be employed as examples of terrorism. All of these events involved the use of force or violence against persons or property in violation of criminal laws of the time. Some would argue that all of these events were attempts to intimidate, coerce, or financially harm legitimate government powers of the time.

In defining terrorism, many people have cited the common idiom "*One man's terrorist is another man's freedom fighter.*" All of the events mentioned above involved both use of force and direct or indirect attacks on legitimate governments, yet it is highly unlikely that many historians would label them as "terrorist acts." Defining terrorism as FEMA has done leaves us with many gaps in our understanding of nontraditional perspectives on terrorism—including, but not limited to, the definitions of terrorism that have been utilized by several other countries on the floor of the United Nations. Some countries have called for open dialogue to understand the causes of certain events before labeling them as terrorism. So if it is difficult to define terrorism, then defining counterterrorism is even more difficult. Therefore, it becomes impossible to distinguish counterterrorism measures, as they are labeled by government agencies, from other acts of law enforcement at the definitional level.

Counterterrorism, for all practical purposes, involves the intervention of government agencies into plots intended to destabilize the authority of a central government. Whereas espionage is a component of counterterrorism, so, too, are direct attacks on perceived threats to the government. In recent years, there have been many attempts to put together a comprehensive history of terrorism, but no attempts at a comprehensive history of counterterrorism have been undertaken. Counterterrorism, therefore, must involve, first and foremost, a study of the way that agencies respond to perceived terrorist threat.

Although it is common for many historians to list the first attempts at what is defined as terrorism to the Roman occupation of Jerusalem (Faulkner, 2002), the idea of studying systemic attempts at counterterrorism is in its embryonic stages. We accept the FEMA definition of terrorism, for practical purposes, but it is also important to understand that counterterrorism, like terrorism, is partially symbolic in nature. Understanding how terrorism is defined, and the actions that government agencies take against certain activities, helps us to understand what events are important enough to the public sentiment to arouse interest or panic to the point where they can accurately and reasonably be labeled as acts of terrorism.

In understanding counterterrorism, it is important to look at the foundational attempts of the League of Nations and its successor, the United Nations, to address terrorism systemically on a global level. During the aftermath of World War I, in 1920, the League of Nations was established as the first organization charged with maintaining international peace. One of the most important arenas that the League of Nations was charged with protecting was the Balkan states; Serbian Archduke Franz Ferdinand was assassinated in 1914 to contribute to the start of World War I (Saul, 2006).

In 1938, shortly before the onset of World War II in Europe, the League of Nations released the first legislative statement on counterterrorism. The release included a document titled "Proceedings of the International Conference on the Repression of Terrorism" and suggested that many counterterrorist measures—such as creating an international port out of a disputed zone in Memel (Klaipeda), Lithuania—be taken to avoid terrorism that could lead to another war in Europe. The focus of the League's concern was religious and political turmoil in Eastern Europe. The League of Nations failed in its attempts to combat terrorism on a global level, and the United Nations suffers today from many of the same problems that plagued the League.

Barring those events that are perhaps better understood as legitimate acts in American history, such as the Confederate invasion of Fort Sumter in 1860, some have argued that the first domestic attempt at counterterrorism in the United States occurred in response to what is now known as the Haymarket Affair. This involved a series of labor disputes that culminated in 1886, when a bomb was tossed into a police line during an at first seemingly peaceful labor rally in Chicago's Haymarket Square. Anarchists took credit for killing eight police officers. Modern counterterrorism in the United States technically began when eight people thought responsible for the Haymarket bombing, including five German immigrants, were arrested by local police. Although the prosecution had very little evidence directly linking the group on trial to the bombing, the jury returned eight guilty verdicts. Popular media accounts labeled the Haymarket Square bombing to be the work of "labor radicals," and there was much nativist response, whereby the citizenry faulted immigrants for many social ills. The Supreme Court refused an appeal on the decision, by which five of the eight men were sentenced to death. These men were put to death by hanging, and the real identity of the bomber was never discovered.

Another important early event in the modern history of domestic counterterrorism involved the assassination of William McKinley at the hands of Leon Czolgosz, a disgruntled ex-employee of the American Steel and Wire Company who was fired after participating in a labor strike. Czolgosz, a Polish immigrant, managed to fire two shots into McKinley at close range in September of 1901. Immediately after the shooting, he was beaten by onlookers as well as by McKinley's security detail; however, he lived to stand trial. Refusing any legal representation, Czolgosz was sentenced to death by a special jury. Before his execution, Czolgosz declared that he had assassinated the president for the well-being of the American people. Lithuanian-born anarchist Emma Goldman was taken into custody shortly after McKinley was shot; however, there was not enough evidence linking her with the assassination.

The McKinley assassination set off a wave of panic about terrorism, and there was an immediate public outcry against both immigrants and unions. The effects of the McKinley assassination colored America's perception of immigrants and unions for years to come. It can be argued that this fear of endangerment at the hands of foreign conspirators agitated the minds of the American public until the end of the McCarthy hearings during the late 1950s. Furthermore, the incident gives leverage to the statement that immigrants have long been burdened with the suspicion of terrorist involvement. In some ways, the reaction of America (both citizens and government agencies) to the McKinley assassination typifies the way that conterterrorist measures arise in response to public sentiment.

Public reaction to the McKinley assassination also seems to underscore a very important point regarding the relationship between agency action and nativist appeal when it comes to making policy decisions regarding counterterrorism. The McKinley assassination, which was followed by many attempted bombings, drew the ire of the American mainstream toward the immigrant population. The surge of immigrants into the United States between the years 1900 and 1910 had swelled to a high of 8 million, an influx that has yet to be surpassed in American history. During the early 1900s, when bombs were anonymously mailed to governmental officials, public sentiment generated a large degree of nativist campaigns for immigration restriction. These campaigns openly sought to achieve the barring of immigration from groups that were suspected of harboring radical ideas. Specifically, Italians, Russians, and Eastern European Jews were all viewed as potential terrorists.

The first systemic attempt at counterterrorism in the United States by any federal agency was when the U.S. Department of Justice began to conduct what are now known as the Palmer Raids. Between the years 1919 and 1921, both the U.S. Justice and Immigration Departments targeted the radical left in the United States. The raids, named for Alexander Mitchell Palmer, who at the time was serving as United States attorney general, resulted in the arrest of over 200 suspected radicals of Russian descent in 1919. Many of these suspected radicals were deported without trial. Ironically, these deportations were clearly a violation of the Immigration Act of 1917, which—prior to the 2001 passage of the USA PATRIOT (Uniting and Strengthening America by Providing Appropriate Tools Required to Intercept and Obstruct Terrorism) Act—required a hearing before deportation regardless of circumstances. In 1920, 6,000 suspected radicals, mostly union members, were rounded up and treated in similar fashion. The nativist and public reaction to these raids was largely positive, and the raids themselves culminated in the Immigration Act of 1924.

Although some would argue that the persecution of American labor union members ended for good with the Roosevelt administration's New Deal and the coming of the Great Depression, the Red Scare—as it has become known to historians—had quite a bit of staying power. However, by 1950, Communists in the United States would no longer be thought of as a terrorist threat, but instead would be seen as an espionage threat to national security.

During the mid-20th century, J. Edgar Hoover, then head of the Federal Bureau of Investigation, pursued inroads into domestic counterterrorism with a new breed of enemy in sight: the Ku Klux Klan. In September of 1963, a bomb exploded at a Baptist church in Alabama, and four Klansmen were accused of the crime. This was terrorism on a new level. Hoover targeted the Klan in the South and throughout the 1960s; domestic counterterrorism was marked by the infiltration of the FBI into the Klan. However, although Hoover's attempted infiltration into the Ku Klux Klan did not forever stop events such as church bombings or the beating of several freedom riders, the fear of infiltration by FBI agents was enough to dissuade many sympathizers from active membership.

With the Cold War and the specter of nuclear extinction looming in the background, domestic terrorism in the United States throughout the

1960s and 1970s brought about only minuscule cause for social concern. Civil defense programs flourished, and the Central Intelligence Agency concentrated its efforts on counterespionage efforts to combat Soviet spies or assassins, but agencies were limited to reactive policy making revolving around domestic terrorism. In 1970, four antiwar protesters exploded a bomb on the University of Wisconsin campus in protest of the war in Vietnam. In December of 1975, the bombing of a TWA terminal at New York's LaGuardia Airport resulted in 11 people killed, yet there were no arrests made. Counterterrorist efforts were at this time poorly organized and, in hindsight, appear as little more than an afterthought.

During the early part of the 1980s, governmental responses remained at best uneven, yet there is evidence of recognition of potential threats associated with international terrorism linked to increased activity by Puerto Rican independent militias within the United States. In 1981, the Puerto Rican Armed Assistance movement destroyed the men's restroom at New York's Kennedy Airport. Again, no arrests were made. In 1983 when the U.S. Capitol was bombed, the threat of terrorism from Puerto Rican terrorist groups briefly captured the public's attention, but not for very long. Although there were no injuries, a wall in the Senate chamber was damaged, and two radicals were arrested. Increased security measures were taken to protect Congress, the Pentagon, and other government buildings at the time, but the incidents did not arouse enough attention to generate any substantial counterterrorism legislation or formation of any auxiliary agencies. The Cold War emphasis took precedence over other threats to American national security.

During the late 1980s and early 1990s, the focus of counterterrorism switched to environmentalist groups such as Greenpeace and the Animal Liberation Front. Again, although there was increased agency activity in pursuing environmentalist terrorist groups, there was no substantial legislation passed or agencies formed to combat the perceived problem.

Modern counterterrorism efforts began to take shape in the early and mid-1990s. Following the first Gulf War with Iraq, the juxtaposition of modern terrorism with Islamic radicalism came into the public eye. On February 26, 1993, Islamic terrorists were arrested and eventually convicted for the bombing of the World Trade Center, where a handful of people were killed and over 1,000 people injured. Popular media accounts of the incident began to relay fears of Saddam Hussein's retaliation after the first Gulf War.

Ironically, although the first attack on the World Trade Center did serve to shift public attention toward Islamic terrorists, the deadliest terrorist attack on American soil prior to 2001 was perpetrated by an American, Timothy McVeigh. On April 19, 1995, McVeigh bombed the Murrah Federal Building in downtown Oklahoma City, killing 168 Americans. Still, there remained the American propensity to find cause with Islamic terrorists, as many popular media accounts shortly following the McVeigh bombing attributed the source of the bombing to Islamic radicals.

It was not until immediately after the McVeigh attack that any substantial legislation was passed to combat terrorism. Although by 1996 the work of McVeigh was known to be associated with certain militant groups and not of Islamic origin, several counterterrorist efforts passed through Congress targeting foreigners. Perhaps due to the newly found vulnerability in America's infrastructure that was exposed by McVeigh—or perhaps due to much public sentiment regarding fears of illegal immigrants—the Antiterrorism and Effective Death Penalty Act of 1996 (AEDPA) and the Illegal Immigration Reform and Immigrant Responsibility Act of 1996 (IIRAIRA) were passed by the Clinton administration. These particular bodies of legislation increased the criminalization of aliens, making grounds for deportation retroactive based on previous crimes committed by immigrants. McVeigh himself was homegrown and did not have a criminal record. Yet after the Oklahoma City bombing, the Immigration and Naturalization Service (INS) rapidly increased alien deportations (Miller, 2005). The noteworthy impact of these laws is that they constructed a new rift between citizens and noncitizens, and deprived the latter of certain constitutional rights that were associated with due process. A notable right that remained intact was the right to be treated fairly in an *open* deportation hearing or a criminal court proceeding.

Since the September 11 terrorist incident, the ideas of terrorism and of counterterrorism have disseminated to an alarmed American public and warrant their own historical account. Before this decade, terrorism in the United States had never been seen at a level where thousands of people could perish from attacks. The scale of the 9/11 disaster brought about many changes with regard to efforts to combat terrorism. Such changes include the creation of the Department of Homeland Security, the passage of the PATRIOT Act, and the passage of the 2006 Military Commissions Act. Also of note is the marked increase of security in public places, particularly those regarding transportation, as well as the lack of scrutiny allotted to paramilitary nativist militia groups, such as the Minutemen. In response to percieved government ineffectiveness in controlling undocumented entries to the United States, these citizen patrols have ranged from neighborhood watch types reporting suspicious crossings to untrained individuals acting as more of a hindrance to the border patrol than any actual assistance. Although it is clear that some people are crossing illegally, the symbolic nature of a nongovernmentally sanctioned group actively patrolling U.S. borders sends a message that all who cross are suspect.

The history of counterterrorism has continued to develop throughout this decade. And with the increase in its prominence, there becomes a need to understand its history.

PRE-9/11 COUNTERTERRORISM: SYMBOLISM AND MORAL PANIC

Stone (2004) describes how United States history has continuously had a cycle of free speech and human rights legislation being created in order to address a problem, implemented against perceived threats to the government, and then ignored after the implementation was seen as infringing on human rights—only to repeat this same process. He lists a series of governmental actions in response to crises, including (p. 12–13):

- The Sedition Act of 1798
- Abraham Lincoln's suspending the writ of habeas corpus
- World War I era: Espionage Act of 1917 and Sedition Act of 1918
- World War II: Japanese internment
- Cold War: McCarthyism
- Vietnam Era: FBI countermovements to limit dissidents

All of these actions placed some limits on speech and opposition to the government. Of course, there is a significant need for security in times of national vulnerability. However, the concern lies in the continuum of low-risk actions to high-risk actions. The nation balances free speech with concerns about inciting a mob or organizing terrorist activity. The solution to this may sound simple. Let the patriotic good citizens speak while limiting the speech of those that mean to do harm to the nation. The question then is: How do we tell the "safe" people from the "threatening" people?

It is important to understand that historically, counterterrorism measures have not been enacted by government agencies without a certain level of public support. To be sure, the nature of terrorism is highly symbolic, and as such, fear of terrorism brings out salient responses in people. Historically, immigrant crime on American soil has always drawn ire from the citizenry, and often has resulted in escalations of violence. A key historical example of the violent response that immigrant crime can often draw is the case of Italian lynchings in Louisiana in 1891. Immigrant crime, especially terrorism, seems to have much populist appeal, which, in turn, presents loud calls for rapid response—calls that trump rational policy study. At many points in U.S. history, the "terror threat" has stepped to the forefront of public concern, and immigrant scapegoating for economic problems has repeatedly occurred (Johnson, 2004). A good example of such a point in U.S. history is the beginning of the 20th century, when the immigration of Eastern and Southern Europeans into the United States totaled over 8 million (U.S. Bureau of Statistics, 1901). A moral panic about the influx of immigrants permeated dominant American culture and resulted in the Immigration Act of 1924 and, arguably, Prohibition. Such moral panics are symbolic in origin and always are reliant upon people's need for social order.

During the early and mid-1990s, a moral panic about the economic, demographic, criminal, and terrorist consequences of loosely regulated immigration standards resulted in a complex stereotyping of immigrants and gave way to a popular view of immigration as a social problem. Cohen (1972, p. 9) identified the concept of moral panic as follows: "A condition, episode, person or persons emerges to become defined as a threat to social values and interest; its nature is presented in a stylized and stereotypical fashion by the mass media and politicians." It may perhaps be of some surprise to the reader, considering the McVeigh attack, that militia and survivalists groups such as the Branch Davidians did not become the center of social policy attention. Throughout history, we have seen many of the values espoused by such militia movements also espoused by the public. Xenophobia, fear of illegal aliens, and the despising of relinquishing the American way of life have been used as justifications for acting against the government. In fact, during the radical labor scare of the early 1900s, many nationalist groups were involved in the lynching and prosecution of immigrants who were perceived as terrorist threats.

Cohen (1972) indicates that moral panic generates public reactions that are out of proportion to the real and present danger of a threat that actually exists. In response to this exaggerated concern, "Folk devils are created, deviant stereotypes identifying the enemy, the source of the threat, selfish, evil wrongdoers who are responsible for the trouble" (Goode & Ben-Yehuda, 1994b, p. 156).

Goode and Ben-Yehuda continue to expand on the idea by listing crucial elements of a moral panic. We will discuss only the linkage between two of them: (1) concern about the behavior of the targeted group; and (2) hostility toward the people seen as engaging in that behavior. One of the questions that comes to light is the linkage between concern about the behavior of immigrants and the level of hostility toward immigrants.

In 1995, the conviction of Muslim cleric Omar Abdel Rahman and nine of his followers increased public fear about national security threats posed by Middle Eastern radicals (Reimers, 1998). Initially, the Oklahoma Federal Building bombing, known to have been perpetrated by two American citizens, was considered by the media to be the work of Middle Eastern terrorists (Welch, 2002). This moral panic, compounded with the knowledge of the Muslim attack on the World Trade Center in 1993, was accompanied by the passage of a series of laws designed to combat undocumented immigration, crime, and terrorism (U.S. Congress, House, 1993, 1994a, 1994b; Senate, 1995). It could be argued that the public accepted these laws to augment their sense of symbolic security and because they perceived that citizens would not be dramatically impacted. The folk devil—the illegal immigrant—as identified as connected to economic problems, crime, and terrorism, was targeted. This involved symbolic politics and reactions to specific iconic symbols such as the World Trade Center as a representation of the capitalist system and corresponding American way of life. Although the loss of life was enormous, we cannot neglect the significant symbolism related to the buildings that fell on 9 /11. The Oklahoma City bombing was clearly a symbolic attack, but it was difficult for the American public to define the threat of a U.S. citizen and military victim as fitting into the category of a folk devil.

Therefore, it is of no surprise that the counterterrorist reaction to the bombing of the Oklahoma City Federal Building involved legislation against foreigners, and not malicious Americans. The AEDPA and IIRAIRA increased the criminalization of aliens, making crimes considered grounds for deportation retroactive, and the Immigration and Naturalization Service rapidly increased deportations (Miller, 2005). These laws drew a sharp line between citizens and noncitizens, and deprived the latter of constitutional rights—a move that was later questioned and has been legally challenged from 2001 to 2006. In America before the 1990s, immigrants were allowed broad constitutional protections, the most important of which was due process.[1] This included the right to be treated fairly in an *open* deportation hearing or a criminal court proceeding.

Subsequent to this legislation, the nation experienced economic prosperity, sentiment regarding legal immigration became more positive, and only negative attitudes toward undocumented immigrants were focused upon (Welch, 2002). In part, this is due to the nation's reliance on skilled legal immigrant labor and undocumented labor in such areas as agriculture. Employers, not workers, want immigrant labor, and workers are less likely to complain when jobs are plentiful; however, fear and loathing of immigrants during these times were still around, although limited to hate groups such as the Aryan Nations. As a result, the volatility of this issue declined until reignited by the intense fear caused by the 9/11 attack. This renewed the moral panic regarding immigration and firmly embedded it in issues of national security.

In developing a field of counterterrorism, a cost/benefit analysis of the influence of counterterrorist policy on immigration may help to further support the idea that many Americans are willing to compromise fiscal austerity for a sense of security. Checkpoints that are set up to stop illegal

---

1. Post-9/11 policies are particularly restrictive for immigrants because of overwhelming public support. However, some would argue that 1996 antiterrorist legislation, which was passed in reaction to the Oklahoma City bombing, had deprived immigrants of due process in a criminal trial five years prior to 9/11 (Welch, 2002, 2003; Foxell, 2004; Hornqvist, 2004; Miller, 2005).

drug flow and terrorists are, and previously were, catching many of the illegal aliens that provide labor for many construction projects (Diaz, 2005) and agricultural operations (Jordan, 2005). Increasingly, undocumented immigrants in search of work are driven to cross borders in dangerously remote arid areas, resulting in the loss of many lives (Eschbach et al., 1999; Cornelius, 2001; Nevins, 2002), causing Inda (2006, p. 165) to refer to the U.S.-Mexican border region as a "landscape of death."

Curtailment of immigrants' rights to due process provides a measure of symbolic security for many citizens. Yet in deference to the perceived security argument, we would also point out that curtailing due process for immigrants provides little to no security improvement and ignores employers' demand for their labor. In border control measures, for example, there is still inadequate staffing and technology as well as flawed visa and biometric systems. Undocumented immigrants other than Mexicans (OTMs) awaiting trial are released into the general U.S. population; this is often referred to as the "catch and release" program (Martin, 2005). Current attempts to stop the entrance of immigrants will do very little toward the provision of homeland security. Such efforts do, however, address the anger of citizens whose desire for security sways their abilities to support symbolic solutions, while still allowing many of those illegal immigrants who are the targets of symbolic efforts to cross U.S. borders in search of employment.

POST-9/11 COUNTERTERRORISM: COUNTERTERRORIST POLICIES AND PERCEIVED SECURITY

To keep social order, people need to "feel" protected from terrorist attacks whether or not a threat of great magnitude exists. This is understandable because, until the first and second World Trade Center attacks, dominant American culture embraced the concept of never having been attacked on the American continent as a unifying point in collective sentiment and tradition. On 9/11, symbols of security and national strength (the Pentagon; possibly Washington, DC; and/or the White House) and pride (the Twin Towers) were targeted by Middle Eastern terrorists in a challenge to U.S. global political and economic power. A fearful public reaction occurred, and the subsequent volatility of moral panic rapidly produced popular sentiment for policy changes. People rallied around aggressive policy making to regain a sense of social cohesion and personal security.

The USA PATRIOT Act was passed at a time when moral panic generated by perceived foreign involvement in terrorism took precedence over concerns for civil liberties. Collective responses to these dramatic events resulted in widespread clamoring for terror protection and produced quick and unobstructed passage of draconian measures to safeguard national security. Forty-five days after September 11, 2001, the USA PATRIOT Act was hurried through Congress with little to no debate. The Homeland Security Act of 2002, which also proposed to provide better security for Americans by reorganizing several key government agencies under the Department of Homeland Security (DHS), also was passed less than a year after the attacks. Much of the scholarly criticism of post-9/11 policies has revolved around the idea that American citizens sacrificed some civil rights for security from terrorism (Foxell, 2004; Miller, 2005; Thornburgh, 2005; Welch, 2002, 2003).

The most concrete national compromise for security amenable to scientific analysis is of course fiscal—and the global "War on Terror" is expensive. Funds allocated to the Department of Homeland Security totaled $27 billion in 2005, with the bulk of that going to the routine operations of agencies such as Customs (U.S. Government FY 2005 Estimated Budget). In early 2003, the Federal Emergency Management Agency (FEMA) also became a small part of the U.S. Department of Homeland Security. FEMA's role is recognized as an effort to prepare the nation for hazards and to competently administer federal response and recovery efforts following any national incident.

Whereas there is only limited evidence that increased spending on counterterrorism has actually deterred any terrorist attacks on America, there is some evidence that policies and institutions formed as counterterrorist measures have elevated levels of bureaucratic inefficiency, as demonstrated in response to *natural* disasters such as Hurricane Katrina, to unparalleled heights. Thus, although policies such as the USA PATRIOT and Homeland Security Acts might create heightened feelings of *perceived* security in segments of the population (although this has not been tested either), whether *basic* securities are augmented by such acts is unclear. Hence, aggressive social policy making revolving around antiterrorism may provide a sense of security for many people, but it does not appear to work well in a market economy that needs to protect itself from natural hazards, large deficits, and inflation.

POST-9/11 COUNTERTERRORISM: SYMBOLIC SECURITY AND THE DUE-PROCESS TRADE-OFF

Why did the American public accept this trade-off? The 9/11 attacks revived the moral panic connected to immigration because of the involvement of Middle Eastern nationals. Immediately after the 9/11 attacks, a CBS poll[2] asked the following question: "*In order to reduce the threat of terrorism, would you be willing or not willing to allow government agencies to monitor the telephone calls and e-mails of ordinary Americans on a regular basis?*" Only 29% were willing, whereas 66% of the respondents said that they were not willing to support those actions by the government. With just a simple change in a few words, the same survey asked: "*In order to reduce the threat of terrorism, would you be willing or not willing to allow government agencies to monitor the telephone calls and e-mails of Americans that the government is suspicious of?*" Over half (56%) of the respondents supported this action, with 4 in 10 (39%) expressing that they were not willing to support those actions from the government. The issue is age-old. We want our civil rights. We want others to have civil rights. But many people support the suspension of those rights when there is a perceived threat. A CNN poll (Williams, 2001) indicated that 45% would support torture for the purpose of gaining information about terrorism. Should we be surprised that torture and harsh treatment have been used on Iraqi prisoners (Taguba, 2004) and are alleged to have occurred in immigrant detention (Welch, 2002; Cole, 2003) in the United States?

For "illegal aliens," the threshold justified for deportation is simply stepping across the border (Miller, 2005). For foreign visa holders and permanent residents, deportation kicks in as paperwork expires, if not renewed. Renewing paperwork is compromised if ethnicity and race, nationality, or religious beliefs place a visitor or immigrant in the social profile of a potential terrorist. The assumption that an immigrant fits a terrorist profile rests on many factors, and unfortunately, as so often is true, application of a working profile is generally better at assisting in ruling out classes of sus-

2. CBS Poll administered April 13–16, 2005 (N=607).

pects than it is at singling out and identifying individual would-be terrorists from the crowd. It is not an exaggeration to say there is a great deal of human error involved in this type of surveillance and investigatory process. As of 2001, America and its agencies and law enforcement have operated under what is known as the 1% doctrine, which promotes agencies' errors toward the side of conservatism in making decisions about who is a terrorist threat and who is not. It is not too large a leap to assume that many agencies act in accord with the 1% doctrine and are more willing to be unforgiving toward immigrants with criminal records or other characteristics that make them seem undesirable.

Many argue that the USA PATRIOT Act crossed the social justice line in depriving citizens of rights to due process, although the public is often not aware of the extent to which a particular practice makes them vulnerable. The act's authorization of secret searches and wiretaps in criminal investigations without probable cause also violates the Fourth Amendment to the Constitution's protection of privacy. Intelligence gathered is potentially permanently secret because the original affidavit authorizing the surveillance cannot be requested, thus limiting an individual's right to test its legality (Cole, 2003). In another example, citizens can be detained as material witnesses and/or subject to a criminal charge as a pretext (Cole, 2003). The public perceives these directives, however, as primarily directed toward Arab and Muslim immigrants.

Public support for the PATRIOT Act is not only an indicator of some Americans' willingness to compromise their constitutional liberties for perceived security, but also a sign that many are willing to sacrifice social justice in support for counterterrorist policies. To be sure, the mainstream sentiment toward Arab-Americans and Muslim Americans immediately following 9/11 was fearful at best, and contentious at worst. Federally mandated hate crime statistics indicated that hate crimes increased from 28 in 2000 to 481 in 2001—a 1,700 % increase after 9/11, including several murders (Human Rights News, 2002). Muslim and Arab organizations received over 2,000 reports of September 11–related bias crimes. Again, militia groups and survivalists, such as those responsible for the attack on the Oklahoma City Federal Building, appear to not arouse public sentiment in the way external threats have historically, and do today. In fact, Minutemen groups openly operate at present along the United States border, with very little intervention from the federal government, and have gained support from the average citizen to end the "threat" of terrorism coming from over the border.

Related to the foregoing is the insufferable affront to social justice represented by the practice of some politicians who take full advantage of mainstream Anti-Arabic sentiment. In this respect, the myths of the War on Terror has paralleled the War on Drugs (Schuck, 2005; Thoumi, 2005) and the War on Crime (Miller, 2005). Politicians understand counterterrorism as a popular political platform. John Cooksey (R-Louisiana), a member of the U.S. House of Representatives, and at that time a member of the House Committee on International Relations' Subcommittee on the Middle East and South Asia, had remarked in a national radio address on September 17, 2001, that "someone come in that's got a diaper on his head and a fan belt wrapped around the diaper on his head, that guy needs to be pulled over."

During the 2004 presidential elections, even the Democratic Party showed no opposition to the general concept of a "War on Terror." Detained immigrants of nationalities associated in the public mind with terrorist threats are highly visible to the public and serve to reinforce symbolic security. This is a carryover from previous symbolic but highly inefficient efforts to "control the border"—efforts that target this group while simultaneously allowing these highly sought after undocumented workers to enter (Andreas, 2000).

The magnitude of U.S. response to the terrorist threat—which includes the wars in Afghanistan and Iraq and dispersed terrorist activity such as the Iraq insurgency, the 2005 London subway suicide bombings, and the strike on three American identified multinational hotel chains in Jordan (Slackman et al., 2005)—has resulted in continued moral panic, but the public is becoming more disturbed about the contrast between real and perceived security. By 2005, a public opinion poll (N=800 adults nationwide) conducted by the Center for Survey Research and Analysis at the University of Connecticut reported that 64% responded affirmatively to the following question: "*The Patriot Act makes it easier for the federal government to collect information on suspicious Americans. Based on what you have read or heard, do you support or oppose this act?*" Of the respondents, 25% were opposed. In response to the question "*Do you think that the Patriot Act has prevented terrorist attacks in the United States?*" the public was divided, with 44% answering yes and 43% no.

Symbolic security cannot be maintained without concrete evidence of success. In the same poll, 20% thought that the USA PATRIOT Act should be renewed, 57% supported renewal with modifications, and 17% wanted it to expire. Two provisions of the bill that a majority of the public does not support are: (1) requiring banks to turn over financial records without a judge's prior approval (55% opposed); and (2) court ability to authorize searches of American homes without prior notice if it would inversely impact criminal investigation (75% opposed). Notably, these aspects of the act have the greatest potential to negatively impact citizens.

In the absence of major terrorist events in America since 9/11, the Senate is debating "privacy concerns" and the reauthorization of the USA PATRIOT Act. The House and Senate will vote on the future of the USA PATRIOT Act. Permanent enactment impacts (restrictions placed on the FBI to access) private citizens' phones, e-mail, Internet sites, and financial data. Judges could be given the power to reject national security letters permitting secret access. Banks and Internet providers also could challenge such letters in court.

POST-9/11 COUNTERTERRORISM: FALSE POSITIVES, MIDDLE EASTERN IMMIGRANT SWEEPS, AND DETENTION

After 9/11, INS focused on Arab and Muslim immigrants as potential "terrorists" in general, with only a sliver of social protest. Middle Eastern immigrants around the country were subjected to arrest and secret hearings—and, in many cases, prolonged or even indefinite detention at facilities (Parker, 2004; Cole, 2003). In the two months after 9/11, the government was detaining over 1,200 immigrants of this background (Welch, 2002).

Today, we are aware that the vast majority of these detainees are "false positives," a group of people who lost their rights when the government erred on the side of caution (Cole, 2003). As of May 2003, Cole (2003) found that only three of the "suspected terrorists" arrested in the first seven weeks after 9/11, and none of 4,000 arrested since that time, were charged with terrorism. Most of these individuals were found to have immigration violations—particularly overstaying their visas (Welch, 2002). Notoriously, a British citizen, Richard Reid, and a French citizen,

Zacarias Moussaoui, individuals charged with serious terrorist acts, do not fit this profile. Similarly, a majority of those since charged with Al Qaeda related crimes are U.S. citizens, not permanent residents or visitors.

Cole (2003) points out that ethnic, racial, nationality, and religious profiling sweeps are not cost effective. "False positives" can cause investigators to overlook individuals they should have scrutinized. Gladwell (2003) indicates that false positives reduce the likelihood of reacting in time to prevent terrorism. Betts (1978) found that raising the sensitivity of reaction increases false positives. It is perhaps no surprise that the actions of Timothy McVeigh went undetected in 1995. McVeigh was white and non-Islamic and appeared, for all practical purposes, to be a patriotic American.

A more intriguing question regards whether restricting immigration can provide concrete security from terrorists and not just symbolic security. A recent report from the Center for Immigration Studies (Camorota, 2002) analyzed the background of the 48 foreign-born Al Qaeda–linked operatives who took part in terrorism within the United States for the 10 years preceding the September 11th attack. The operatives were classified in various categories by the INS:

- One-third were on temporary (tourist) visas and subject to minimal federal regulation due to the expectation of their short duration of stay in the United States.
- One-third were legal permanent residents or even naturalized citizens.
- One-fourth were illegal aliens and subject to scrutiny by the government for illegal activities.

If we focus just on the terrorists involved in 9/11, we can note that all the hijackers entered the country legally, and that a few suspected potential conspirators were denied visas. Granted, many people argue that had there been a better ability to investigate the backgrounds of the terrorists, the attacks could have been prevented. Government reports released in 2004 have stated that immigration officials failed to recognize problems with the hijackers' passport and visa applications. The failure of the FBI, CIA, and INS to share information also is a notable tragedy. That information, if combined, would have allowed the detection of the Al Qaeda operatives at flight schools and would have predicted that the World Trade Center would be a target.

Basically, because a small number of terrorists were able to enter, the post-9/11 sweeps of Middle Eastern and Muslim immigrants were designed to "find the rest." The leap to making foreign visitors and immigrants responsible for terrorism is like statements claiming the solution is to "close the border." They provide an oversimplified solution to the problem of terrorism in the United States.

The massive number of false positives, injured lives, and continued human suffering of detainees provided "symbolic security" and evidence of government action, but proved a failure and an example of how not to conduct terrorist investigations. Successful terrorist operations involve infiltration of terrorist sleeper cells and the use of paid informants, some of whom provide "dirty" details because they are involved in crime or have been involved in terrorism. Jenkins (2003) points out that this type of government activity is *not* public because it would render the intelligence useless. Instead, DHS publicly pushes operations that attack a social category of individuals and their communities—operations that will destroy trust and result in a break in ties with law enforcement (Cole, 2003).

COUNTERTERRORISM AS OF 2007: EFFECTIVE VS. INEFFECTIVE COUNTERTERRORIST MEASURES

In terms of counterterrorism, the first question that must be answered is: Are terrorist acts crimes, or are they acts of war? This dual definition of the situation leads us to pursue both short-term solutions—such as use of the military, policing, and global intelligence—as well as long-term solutions that call for nation building to stop the development of terrorist networks (Steinberg, 2002).

First of all, investigators must find the party responsible for terrorist actions. This is a complex issue. Who perpetrated the act of terrorism, and who assisted in the action? The degree of liability for each action is subject to debate in any criminal act. A counterterrorism response inside the United States is more complex than calling for an air strike on a foreign country or some other military solution. Frequently, the responsible party cannot be clearly tied to a small geographic location or to a state sponsor.

A second issue develops from the fact that intelligence gathering is a critical aspect in understanding the causes of terrorist acts, the likelihood that harmful actions may take place, as well as the specific details of targeted actions that are being planned. Once the valuable information is discovered, who shares the information with whom? The issue of confidentiality and the secrecy of classified information does not allow for a full-scale public relations barrage, explaining the actual number of terrorist attacks that were prevented; therefore, success and failure are difficult to measure. One of the few measures of success in terms of counterterrorism serving as capable guardians of the state can be the level of trust that citizens place in the counterterrorist agencies.

Finally, we must understand and define a protocol for how information technology and communication among agencies will be shared. This will lead to the next step of judging the quality of the information so that the agencies may make rapid decisions and respond accordingly.

All of this calls for the integration of global intelligence with both local and international policing agencies. The most obvious courses of action related to counterterrorism are (1) prevention; (2) monitoring and banning of groups that encourage violence; and (3) building relationships between nations. These actions are challenging in that they must follow the Constitution of the United States and respect the laws of other national governments, while maintaining effective deterrence and rapid responses to terrorist actions.

CONCLUSIONS

We propose that a history of counterterrorism is the first necessary step toward understanding some of the shortcomings of counterterrorism policies and actions today. Deflem (2004) has proposed a "Sociology of Counterterrorism." Such a field could also assist in understanding some of the causes and consequences of hastily structured policies, such as the USA PATRIOT Act and the Homeland Security Act of 2002, as well as the continuous restructuring of the Department of Homeland Security (DHS). As part of the Homeland Security Act of 2002, the DHS was restructured to coordinate the services of various federal law enforcement divisions such as the Immigration and Naturalization Service (INS) and the U.S. Customs Service. These divisions are entering into cross-pressures in attempts to maintain their bureaucratic structure, mission, and focus while subsequently attempting to appease the political motivations of the American people (Deflem, 2004).

We feel that the time has come for social scientists to recognize the imperative of building an effective sociology of counterterror. The efficiency of having one clear vision and focus leads to efficient police work and service delivery. Post-9/11 policies and institutions have been rapidly combined for efficiency, but as Deflem (2004) points out, each agency and policy leads to a bureaucratic yield so large as to be problematic for law enforcement agencies. The same is true with the political redirection stated in many new homeland security policies. This includes the extraordinary number of false positives produced by DHS immigration sweeps, the disutility of the color-coded warning system, and the lack of attention to basic protective infrastructure for natural disaster.

A sociology of counterterrorism also would aid policy makers in making decisions about where to go from here. *Should we be more careful about who gets a green card and naturalization papers*? Of course, the answer is yes. But then we face the harsh reality that INS is already overburdened and backlogged with numerous cases that have not been processed. *Is the answer to homeland security greater scrutiny at the borders*? Yes, but that is only a small part of the answer. *Is the answer greater scrutiny of those individuals who receive permanent resident status*? Yes, but that is only part of the answer. *Is the answer to homeland security screening potential tourists and visitors*? Yes, but that is only part of the answer to the problem.

In the absence of a comprehensive sociology of counterterrorism, policy makers continue to make counterterrorist legislation that appeals to popular sentiment without academic scrutiny. They are essentially "working in the dark." The goal of an effective counterterrorist policy must be to: (1) accurately detect the presence or absence of terrorist threats with the fewest false positives or false negatives; (2) promote an overall atmosphere that does not degrade the concept of social justice; and (3) recognize that legislation based on strong collective sentiments about symbols tends to produce over-bureaucratization and counterproductive results. Therefore, developing a "Sociology of Counterterrorism" is a logical and necessary initial step to ensure that these three standards are met.

In the absence of a comprehensive sociology of counterterrorism, affronts to social justice against immigrants are too easily overlooked. According to the Urban Institute (Fix & Passel, 1994), U.S. immigration is dictated by five goals: (1) the social goal of family unification; (2) the economic goal of increasing U.S. productivity and standard of living; (3) the cultural goal of promoting diversity; (4) the moral goal of promoting human rights; and (5) the national and economic security goal of preventing illegal immigration. It is widely understood that critics of immigration overlook the noneconomic goals, as well as the long-run economic benefits of immigration.

The symbolic dimensions of policies such as the USA PATRIOT Act are extremely appealing to many of the American public because they are geared to dissect and scrutinize current and potential immigrants, a group that has served as a scapegoat for past economic ills repeatedly in the nation's history (Johnson, 2004). Certainly, the profiling of immigrants during a time of war provides an easy solution to a complex problem. This profiling has precedents during World War II. After the bombing of Pearl Harbor began the U.S. war with Japan, Japanese citizens on the West Coast were placed in internment camps (Takaki, 1994). Detention, such as that practiced on Arab and Muslim Americans after 9/11, presents the public with a perpetual "straw man" and reaffirms symbolic security (Cole, 2003), but has little relevance to meaningful safety and physical security.

Many researchers (Das Gupta, 2004; Lafer, 2004) have looked upon the social injustice of post-9/11 immigration policies as unacceptable, including factors such as the following: (1) the ability of the INS to indefinitely detain an alien if the attorney general decides that the alien poses a significant risk of committing terrorism; (2) secret evidence and classified government documents used as evidence; (3) hearings conducted in secret; and (4) the increased civil and criminal penalties for naturalized immigrants that are an affront to social justice. However, without a unified sociological perspective to indicate why such post-9/11 policies are supported by the public, researchers have been unable to provide suggestions about the predictable future of such policies.

REFERENCES

Andreas, P. (2000). *Border games: Policing the U.S.-Mexico divide*. Ithaca, NY: Cornell University Press.

Betts, R. K. (1978). "Analysis, war and decision: Why intelligence failures are inevitable." *World politics* 31: 61–89.

Camorota, S. (2002). *The open door: How militant terrorists entered and remained in the United States, 1993-2001*. New York: Center for Immigration Studies. Retrieved November 14, 2005, from http://www.cis.org/articles/2002/Paper21/terrorism.html.

Cohen, S. (1972). *Folk devils and moral panics*. London: Macgibbon & Kee.

Cole, D. (2003). *Enemy aliens: Double standards and constitutional freedoms in the war on terrorism*. New York: New Press.

Cornelius, W. A. (2001). "Death at the border: Efficacy and unintended consequences of U.S. immigration control policy." *Population and development review* 27(4): 661–685.

Das Gupta, M. (2004). "A view of post 9/11 justice from below." *Peace review* 16(2): 141–148.

Deflem, M. (2004). "Social control and the policing of terrorism foundations for a sociology of counterterrorism." *American sociologist* 35(2): 75–92.

Diaz, E. (2005, May 7). "Last big issue is migrant hiring." *Arizona Republic* p. A-1.

Durand, J. & Massey, D. S. (2003). "The costs of contradiction: U.S. border policy 1986–2000." *Latino studies* 1(2): 233–252.

Eschbach, K., Hagan, J., Rodriguez, N., Hernandez-Leon, R., & Bailey, S. (1999). "Death at the border." *International migration review* 33(2): 430–454.

Faulkner, N. (2002). "Apocalypse: The great Jewish revolt against Rome, 66–73 C.E." *History Today* 52(10): 47–53.

Fix, M. & Passel, J. (1994). *Immigration and immigrants: Setting the record straight*. Washington, DC: Urban Institute Press.

Foxell, J., Jr. (2004). "United States policy on terrorism: Where are we going and how are we getting there?" *American foreign policy interests* 26(3): 241–252.

Gladwell, M. (2003, March 10). "Connecting the dots." *New Yorker* p. 83, 88.

Goode, E. & Ben-Yehuda, N. (1994a). *Moral panic: The social construction of deviance*. Cambridge, MA: Blackwell.

Goode, E. & Ben-Yehuda, N. (1994b). "Moral panics: Culture, politics, and social construction." *Annual review of sociology* 20: 149–171.

Green, D. M. & Swets, J. A. (1966). *Signal detection theory and psychophysics*. New York: Wiley.

Hornqvist, M. (2004). "Risk assessments and public order disturbances: New European guidelines for the use of force?" *Journal of Scandinavian studies in criminology and crime prevention* 5: 4–26.

Human Rights News. (2002). "U.S. officials should have been better prepared for hate crime wave: Anti-Muslim bias crimes rose 1700 percent after September 11." Retrieved January 10, 2006, from http://hrw.org/press/2002/11/usa-hate.htm.

Inda, J. X. (2006). *Targeting immigrants: Government, technology and ethics*. New York: Blackwell.

Jenkins, P. (2003). *Images of terror: What we can and can't know about terrorism*. New York: Aldine de Gruyter.

Johnson, K. R. (2004). *The "huddled masses" myth: Immigration and civil rights*. Philadelphia: Temple University Press.

Jordan, M. (2005, March 11). "As border tightens, growers see threat to 'Winter salad bowl.'" *Wall Street Journal* p. A1, A8.

Kennedy, J. E. (2002). "Monstrous offenders and the search for solidarity through modern punishment." *Hastings law journal* 51: 829, 832.

Kilburn, J. C., Jr. & Shrum, W. (1998). "Private and collective protection in urban areas." *Urban affairs review* 33: 790–812.

Lafer, G. (2004). "Neoliberalism by other means: The 'war on terror' at home and abroad." *New political science* 26(3): 323–346.

Martin, G. (2005, March 15). "Border still open to terrorists." *San Antonio Express-News* p. 1A, 6A.

McLaren, P. & (2004). "The legend of the Bush gang: Imperialism, war, and propaganda." *Cultural Studies<=>Critical Methodologies* 4(3): 281–303.

Miller, T. A. (2005). "By any means necessary: Collateral civil penalties and the war on terror." In C. Mele & T. A. Miller (Eds.), *Civil penalties, social consequences*. New York: Routledge.

Nevins, J. (2002). *Operation gatekeeper: The rise of the "illegal alien" and the making of the U.S.-Mexico boundary*. New York: Routledge.

Parker, A. (2004, October 1). "Inalienable rights: Can human rights law help to end U.S. mistreatment of non-citizens?" *American Prospect Online Edition*.

Reimers, D. M. (1998). *Unwelcome strangers: American identity and the turn against immigration*. New York: Columbia University Press.

Saul, B. (2006). "The legal response of the League of Nations to terrorism." *Journal of international criminal justice* 4: 78–102.

Schuck, A. M. (2005). "American crime prevention: Trends and new frontiers." *Canadian journal of criminology and criminal justice* 47(2): 447–462.

Shane, S. & Shanker, T. (2005, September 28). "When storm Hit, national guard was deluged too." *New York Times* p. A1, A18.

Slackman, M., Ma'ayeh, S., & Stout, D. (2005, November 9). "Attacks at U.S.-based hotels in Aman were minutes apart." *New York Times* [online]. Retrieved November 11, 2005, from http://www.nytimes.com/2005/11/09/international/middleeast/09cnd-jordan.html.

Sorkin, R. D. & Hays, C. J. (2001). "Signal detection analysis of group decision making." *Psychological review* 108: 183–203.

Steinberg, J. (2002). "Counterterrorism: A new organizing principle for American national security?" *Brookings review* (Summer): 4–7.

Stone, G. R. (2004). *Perilous times: Free speech in wartime*. New York: Norton.

Taguba, A. M. (2004). *U.S. Army 15-6 report of abuse of prisoners in Iraq*. Retrieved November 14, 2005, from http://en.wikisource.org/wiki/US_Army_15-6_Report_of_Abuse_of_Prisoners_in_Iraq.

Takaki, R. T. (1994). *A different mirror: A history of multicultural America*. New York: Back Bay Books.

Thornburgh, D. (2005). "Balancing civil liberties and homeland security: Does the USA Patriot Act avoid Justice Robert H. Jackson's 'suicide pact'?" *Albany law review* 68(4): 801–813.

Thoumi, F. E. (2005). "The Colombian competitive advantage in illegal drugs. The role of the policies and institutional changes." *Journal of drug issues* 35(1): 7–25.

U.S. Bureau of Statistics. (1901). *Statistical abstract of the United States: 1900.* Washington, DC: U.S. Government Printing Office.

U.S. Congress. House. (1993). *World Trade Center bombing: Terror hits home.* Hearing before the Subcommittee on Crime and Criminal Justice of the Committee on the Judiciary. 103rd Cong., 1st sess. Washington DC: U.S. Government Printing Office.

U.S. Congress. House. (1994a). *Criminal aliens.* Hearing before the Subcommittee on International Law, Immigration and Refugees of the Committee on the Judiciary. 103rd Cong., 1st sess. Washington DC: U.S. Government Printing Office.

U.S. Congress. House. (1994b). *Border violence.* Hearing before the Subcommittee on International Law, Immigration and Refugees of the Committee on the Judiciary. 103rd Cong., 1st sess. Washington DC: U.S. Government Printing Office.

U.S. Congress. Senate. (1995). *Criminal aliens in the United States.* Senate Report 104-108, Committee on Governmental Affairs. 104th Cong., 1st sess. Washington DC: U.S. Government Printing Office.

Welch, M. (2002). *Detained: Immigration laws and the expanding I.N.S. jail complex.* Philadelphia: Temple University Press.

Welch, M. (2003). "Ironies of social Control and the Criminalization of Immigrants." *Crime, law and social change* 39: 319–337.

Williams, P. (2001). "This dangerous patriot's game." *Observer.* Retrieved January 10, 2006, from http://observer.guardian.co.uk/libertywatch/story/0,1373,610367,00.html.

CHAPTER 6

Fear of Terrorism

Ashley Nellis

INTRODUCTION

It has been more than five years since the deadliest terrorist attacks in America, leaving our sense of security shaken. Early reports claimed that Americans were quite fearful of terrorism, but fear has waned somewhat as time distances us from the attacks. Few studies of the ways in which emotions, perceptions, and behaviors change in response to terrorism-related information have been published. Specifically, fear of terrorism research seeks to uncover how the American public copes with information about terrorist attacks, threats of terrorist attacks, and antiterrorism policy initiatives that are delivered by the government principally through the media.

Terrorism is a relatively new social problem in America. As we go about our daily activities, there are new worries on our minds about threats over which we have little control. The field of social science is duty-bound to explore the impact of terrorism attacks *and* news of threats of attacks in the citizenry. This chapter reviews results of several important studies in the fields of crime, media, and terrorism research.

RELEVANT RESEARCH

A sound framework for studying fear of terrorism can be found in the crime literature, where it has been studied extensively since the mid-1960s. It is important to explore whether important patterns found among fear of crime studies persist for fear of terrorism because crime fear has been found to be associated with changes in attitudes, cognitions, behaviors, and policy preferences. Two areas of research within the literature on fear of crime are of particular interest and are discussed in this chapter: the role of the media in affecting fear of terrorism, and the role of fear of terrorism in affecting support for certain types of antiterrorism policies.

The media's treatment of crime holds many similarities to its treatment of terrorism (Altheide, 2002). Because of the unique position of the media to inform individuals about terrorism and threats of terrorism, a research concern is the impact of media consumption on fear of terrorism, specifically emotional (fear) and cognitive (perceived risk) responses to terrorism.

Fear is a powerful response to terrorism that may be distorted through media images and messages (Altheide, 2002, 2004). An overly fearful public with distorted risk perceptions may endorse different policies than would a public with objective assessment of terrorism risk. Findings on the relationship between media, fear, and perceived risk, as well as the role of fear and risk perception in antiterrorism policy support, should be important to government officials interested in using the media as an information and public education tool for concerned citizens. Findings should be important to anyone in the academic community who is interested in knowing the sustainability of fear of crime findings in other domains. Finally, findings should be important to members of the public as they learn to cope with terrorism-related information in this post-9/11 era.

The study of reactions to terrorism-related information requires a multidisciplinary approach. Several disciplines—including criminology, political science, psychology, social psychology, risk and decision making, media communications, and sociology—contribute significantly to an understanding of how the public copes with terrorism-related information.

Fear of Crime

Fear of crime is typically defined as the emotional responses of anxiety or dread associated with potential victimization (Ferraro, 1995). The fear of crime literature can be divided into two broad categories of analysis: individual-level correlates of fear of crime, and community-level correlates of fear of crime. Fear of crime is correlated with a number of individual-level factors. The strongest and most consistent predictors of fear appear to be gender (Ferraro, 1995; Fisher & Sloan, 2003; Madriz, 1997; Warr, 1984), age (Ferraro & LaGrange, 1992; Ferraro, 1995; Tulloch, 2000; Warr, 1984), and race (Chiricos & Eschholz, 2002; Chiricos, McEntire, &Gertz, 2001; Covington & Taylor, 1991; Skogan, 1995). Other correlates of fear sometimes include victimization, income, and education, though comparatively little research has been published on the direction and strength of these relationships. Research on community-level correlates of fear considers the impact of physical preservation of an area (including upkeep of buildings, absence of homelessness and public drunkenness, and adequate lighting on streets) and social integration (such as acquaintanceship with neighbors and willingness to be out at night) on fear of crime.

Turning first to individual-level indicators, research consistently finds women to be more fearful about crime than are men. Kenneth Ferraro (1995) clarified our understanding of female fear through his pioneering national study that considered the impact of offense-specific fears rather than the traditional global measures that had typically been used to measure fear of crime. With respect to fear of crime among females, this proved to be especially enlightening. The introduction of offense-specific questioning about crimes revealed that women's fear of crime could more accurately be defined as a heightened fear of sexual assault (Ferraro, 1995).

Other reasons for women's elevated fears have been explained in the literature as well. Hale (1996) reports two general explanations. First, both official and unofficial victimization data fail to capture the full nature of female victimization. This may be because crimes that are more commonly carried out against women (e.g., rape and domestic violence) are more likely to be unreported. A second reason may be that women are more commonly subjected to a subcriminal type of victimization categorized by "'hey honey' harassments" that produce fear. Scholars argue that if these types of victimizations were included in overall victimization rates, the fear of crime would not appear to be disproportionate to their rates of victimization (Sacco, 1995; Tulloch, 2000).

Another explanation for higher fear of crime among women is that women may feel more physically and/or socially vulnerable, which translates into higher perceptions of risk (Warr, 1994; Warr & Stafford,

1983). Skogan and Maxfield (1981), suggesting that physical vulnerability refers to openness to attack, powerlessness to resist attack, and exposure to traumatic physical and emotional consequences if attacked. Social vulnerability refers to the concept that some groups are more vulnerable to victimization because of their social standing. Social status—being female, in this case—may cause women to feel more susceptible to victimization.

Race is another common correlate of fear of victimization, and researchers studying the relationships between race and fear typically ask one of two research questions. The first question is whether different races experience different levels of fear of crime. Warr (1994), for instance, finds that African Americans are substantially more fearful of crime than are white people (see also Skogan, 1995). Unlike differences across ages and between genders, racial differences with respect to fear of crime are quite consistent with objective risk; African Americans are more likely to live in high-crime neighborhoods than are whites (Skogan & Maxfield, 1981). In contrast, Taylor's (2001) research shows that compared to whites, African Americans routinely report higher levels of fear, even after controlling for features of the neighborhood such as crime, race, and the presence of social incivilities.

The research on race and fear of crime has also begun to consider races other than African American and Caucasian, as well as other ethnicities (e.g., Chiricos, McEntire, & Gertz, 2001; Lane & Meeker, 2003; Lee & Ulmer, 2000); some evidence shows that Latinos may be even more fearful of crime than are African Americans (Chiricos, McEntire, & Gertz, 2001).

Until the early 1980s, researchers believed that age was positively related to fear of crime such that as people aged, they become more afraid of crime (Clemente & Kleiman, 1979; Skogan & Maxfield, 1981). The elderly had become "prisoners in their own homes" because of their fear of victimization. More recent studies contradict these early findings; the relationship between age and fear is more complicated than previously understood (Ferraro, 1995; Ferraro & LaGrange, 1992; Rountree & Land, 1996). When separated by offense, findings generally show that the elderly are not statistically more fearful of crime. In fact, Ferraro's research (1995) reports that the only victimization-related activity in which older people reported more fear was being approached by a beggar/panhandler; among these, only older women were more afraid.

Community-level correlates of fear of crime consider the impact of physical and social disorder as contributing factors to fear. Physical disorder is marked by litter, abandoned buildings and cars, graffiti, broken and barricaded windows, and unkempt lots. Social disorder is characterized by disruptive social behavior such as rowdy youths, homelessness, drunkenness, and beggars (Garofalo & Laub, 1978; Kelling & Coles, 1996; Lewis & Salem, 1986; Skogan, 1990; Taylor, 2001; Wilson & Kelling, 1982).

Research on the community-level correlates of fear is situated largely in Social Disorganization Theory. Attributed to the 1950s work of Chicago sociologists Shaw and McKay (1969), this theory argues that urbanization and industrialization lead to pockets of low socioeconomic areas with high population turnover rates and increased heterogeneity. This in turn leads to breakdowns in neighborhood cohesion, which leads to crime and delinquency. Scholars define social disorganization in various ways, but major components typically include the strength of local friendship networks, the supervision of neighborhood youths, and organizational development (Shaw & McKay, 1969).

Social Disorganization Theory has received renewed interest in recent decades as a framework for studying crime and fear of crime (Kubrin & Weitzer, 2003; Markowitz, Bellair, Liska, & Liu, 2001; Taylor, 2002). The family of models that has emerged has been called incivilities thesis, disorder/decline, and Broken Windows Theory (Lewis & Salem, 1986; Skogan, 1990; Wilson & Kelling, 1982). Each of them tests the notion that neighborhood ecological factors affect crime and crime-related characteristics such as fear of crime. Taylor (2001) suggests that area residents look to environmental cues to calculate their safety, and that they find physical and social disorder to be fear-inducing. Researchers studying fear of crime commonly claim that breakdowns in neighborhood cohesion cause fear and social isolation, which invites crime to one's area (Skogan, 1990). The effect of this is more fear of crime. There is some empirical evidence to support this claim (Taylor, 2001).

The sociological implications of fear of crime may be equally or more important than the incidence of crime itself. Warr (1980, p. 458) suggests that:

> . . . to the extent that public beliefs about crime diverge from objective conditions, those beliefs may have greater sociological significance than crime itself. That is the case because many of the purported conse-

> quences of crime (fear of strangers, outmigration, decreased social solidarity, excessive security expenditures, erosion of norms, poor police/community relations, loss of confidence in the administration of justice) may result, not from crime itself, but from public beliefs about crime.

Empirical research identifies a number of consequences associated with fear of crime, and study findings are useful guides for the current study. Warr (1994) suggests that consequences of fear can be divided into two broad categories: avoidance behaviors and precautionary/protective behaviors. Survey research indicates that avoidance behaviors are the most common responses. They include spatial avoidance as well as temporal avoidance (Warr, 2000). People who are fearful of crime may avoid certain places (e.g., dark alleys) or may avoid activities at certain times (e.g., nighttime) that they deem to be unsafe. Warr (1994) suggests that spatial avoidance is so common that the ecology of U.S. cities is framed by the patterns it creates.

Precautionary and protective behaviors include seeking the company of others, avoidance of certain types of travel (e.g., driving instead of walking), carrying weapons, placing locks on doors, identifying persons before letting them in, stopping mail delivery when out of town so as to deflect attention from an unoccupied home, placing bars on windows, installing car and house alarms, and purchasing theft insurance (Warr, 1994).

Perceived Risk of Victimization

In her review of the fear of crime literature, Hale (1996) states that fear of crime is commonly defined as the "negative emotion" that results from crime or its associated symbols, and that it is conceptually distinct from judgments or risks. By contrast, risk is a cognitive response, or calculated likelihood, of victimization (Ferraro, 1995).

Research on fear of crime has not always included separate measures of perceived risk, although it has became a standard measure in more recent years. Studies usually find that perceived risk is a strong predictor of fear. Lewis and Maxfield (1980) found that social incivilities such as homelessness predict perceived risk of victimization. Similarly, Warr (1980) found that individuals gauge their safety (i.e., risk) by paying attention to physical cues in their environment, and engage in precautionary or protective behaviors in response.

Warr and Stafford's (1983) findings contradict the commonly held assumption that perceived offense seriousness alone predicts fear of crime. In their mail survey of Seattle residents, they asked respondents to report their fear of 16 specific crimes as well as their perceived risk of victimization by these offenses. Respondents reported how serious they considered these offenses to be. The findings led the authors to conclude that fear of victimization is a multiplicative function of perceived risk and perceived offense seriousness because these two factors carried nearly identical weight in their study. In addition, fear is not necessarily highest for violent crimes. Rather, fear is highest for property offenses. They conclude that although risk perception is a good predictor of fear, perceived risk is not perfectly correlated with fear because fear also depends on individual-level perceived seriousness of the offense in question and on an individual's risk sensitivity.

LaGrange, Ferraro, and Supancic (1992) examine the influence of several social and physical incivilities on fear and perceived risk separately in their survey of 1,101 adults nationwide. They find that incivilities influence fear of crime, but that this relationship is mediated by perceived risk of victimization. Ferraro's (1995) national study of emotional and cognitive reactions to possible victimization yields similar findings: individuals were asked a battery of questions about specific offenses. Findings show that perceived risk is a predictor of fear of crime.

Rountree and Land (1996) also examine perceived risk apart from fear of crime. Their telephone survey of 5,090 Seattle residents questioned respondents about their perceived risk of victimization as a possible predictor of fear of crime. However, this study suffers from one possible threat to validity as a result of the operationalization of the perceived risk measure. Perceived risk is measured by asking individuals whether they worry at least once a week that their home will be burglarized, and whether they perceive that their neighborhood is safe from crime. These measures are problematic because they are too vague and they do not sufficiently measure perceived risk.

The vast majority of fear of crime studies identify community perceptions as a key determinant of perceived risk, which in turn predicts fear of crime (Austin, Furr, & Spine, 2002; Chiricos, McEntire, & Gertz, 2001; Kanan & Pruitt, 2002). This focus on community is useful to the crime literature because crime is ultimately a local issue. Terrorism, however, is a national and international issue, and the public cannot rely solely on social and physical cues to determine people's risk. The public derives its terrorism information exclusively from the media, which makes the media the key factor in risk perception.

Over the past three decades, risk scholars including Paul Slovic, Baruch Fischhoff, Amos Tversky, and Daniel Kahneman have sought to explain the seeming illogicality of strong public concerns over relatively minor risks and vice versa. They have explored risk perception in depth, and much of their work is situated within the social amplification of risk framework (Kasperson et al., 1988). They state that perceptions and *misperceptions* of risk, both by the public and government officials, play a pivotal role in the ability to deal with the threat of terrorism (Slovic & Weber, 2003).

Risk research identifies several features of risk that define our perceptions about safety that are sometimes contradictory. For instance, people tend to judge the likelihood of an event to be higher if they have an emotional or vivid memory of it; this judgment bias is called availability heuristics, and involves building a mental model of an event. Second, lack of specific information (i.e., uncertainty) about a possible event is associated with greater perceived risk. "Optimism bias," is a concept referring to the observation that even when individuals believe that the risk of an event is high, many cannot personalize this risk. Therefore, a disparity exists between one's perceived risk of an event's occurring and one's perceived risk of it's affecting them, even though it may be occurring in their general proximity. Finally, newer risks (such as recent warnings about global warming) often result in higher perceptions of risk than their objective risk (Slovic, Fischhoff, & Lichenstein, 1979; Weinstein, 1989).

Risk analysts operating within the social amplification of risk framework (SARF) argue that "... hazardous events interact with psychological, social, institutional, and cultural processes in ways that may amplify or attenuate public responses to the risk or risk event" (Kasperson et al., 1988). The framework suggests that risk perception is the end result of two major stages. First, information is transmitted directly or indirectly, and is influenced by the volume and tone of the message, conflicting information around the message, dramatization around the message, and symbolic connotations of the message (Kasperson et al., 1988). Together, these components work to amplify or attenuate perceived risk. Scholars note the potential that direct exposure to a catastrophic event may have in further amplifying or possibly attenuating the effects of information transmittal (Kasperson et al., 1988).

The second stage of social amplification involves the interpretation and response to information received in the first stage. This phase, too, comprises several elements. Availability heuristics are used to personalize or synthesize the information received. Individuals look to their own experiences and memory for ways to interpret the information. Theorists warn that this may result in biases or distortions that affect their perceptions. People also respond to information through the filter of their political and social group orientation.

Signal value refers to the messages that information receivers try to decipher beneath the messages they hear. For instance, the August 11, 2006, last-minute discovery of liquid explosives onboard a commercial flight headed from Britain to the United States could be interpreted by the public as a failure of the federal government to stay in control of possible terrorist events. If so, this would amplify concerns of future attacks. Finally, theorists argue that stigmatization of an event may result in avoidance of persons or places, which could, in turn, have social and policy-related consequences. This was observed in the days and weeks following September 11, 2001, when many airline passengers canceled their flights. The U.S. Bureau of Transportation Statistics reports a 34% drop in passenger air travel between the months of September 2000 and September 2001 (Bureau of Transportation Statistics, 2007).

Central to the SARF is the notion that secondary or tertiary consequences emanate from an original event, and these consequences increase or decrease the perceived risk associated with the first event. September 11, 2001, offers many examples of rippling effects of the attacks: bankruptcies across the airplane industry, security enhancements, sweeping restrictions on civil liberties, the war in Iraq, and the "War on Terror" are a few examples. According to this model, one's current perceived risk of terrorism incorporates all of the rippling effects of the primary event.

Researchers have refined the SARF over time, and have adapted it to a myriad environmental, technological, terror-, and health-related issues to better understand individual and collective perceptions and reactions (Kasperson et al., 2003; Slovic & Weber, 2003). Components of this framework have been applied to crime research, too, although infrequently

(Feeley & Simon, 1994; Jackson, 2006). Unfortunately, most of this literature remains in the theoretical domain.

One important refinement that is pertinent to the present study is the importance of trust in government and institutions in the calculation of risk. Slovic (2000) posits that trust is much easier to dismantle than it is to build; once it is broken, it is very difficult to regain. Distrust of government may even amplify risk perception (Flynn, Slovic, & Mertz, 1993).

Risk determination may dictate one's willingness to relinquish some civil liberties or advocate for harsher policies. Slovic (1987) notes, "people make quantitative judgments about the current and desired riskiness of diverse hazards and the desired regulation of each" (p. 281). He comments that the following factors mediate the amount of risk the public will allow: whether the risk is voluntary, has some benefit, is familiar; whether one can maintain some control over the outcome; and the potential for catastrophe, equity, and level of knowledge. Slovic presented several hypothetical risks to subjects and then analyzed the results. Findings revealed that the higher one's perceived risk, the more one advocates for stricter regulations to reduce this risk (Slovic, 1987).

The architects of the social amplification of risk framework argue that the concept of perceived risk as a multiplicative function of severity and likelihood is underdeveloped. Perceived risk must be considered within the context in which it is situated because it is affected by many processes, both personal and political. Many factors may work together to amplify or attenuate perceived risks.

News Exposure, Fear, and Perceived Risk

The news media's influence on fear of crime has been examined in depth (Eschholz, 1997; Hale, 1996). Research is guided by the general theoretical claim that people become afraid disproportionately to their objective risk of victimization because they are influenced by the information they receive about crime through informal sources such as the media, peers, family members, or neighbors (Lane & Meeker, 2003; Skogan & Maxfield, 1981; Tyler, 1980). Altheide (2002) observes that media coverage of crime has been steadily increasing, though crime has been declining. He and others surmise that news coverage of crime affects fear of crime, although the dynamics of that relationship remain unclear (Heath & Gilbert, 1996; Warr, 1980).

Some argue that the media's presentation of terrorism taps news viewers who are already drawn to crime news. "Audience familiarity with terrorism traded on decades of news and popular culture depictions of myths about the 'crime problem,' crime victims, and the drug war" (Altheide, 2004). Media hype about terrorism may inflate fears or perceptions of risk of a terrorist attack. There is some evidence of this in the fear of crime research.

News Types

Research in the communications literature explores differences in the impact of news on emotions and perceptions when it is delivered through print versus audiovisual footage. There are two main differences between these media types, broadly labeled as content differences and consumer selectivity differences. First, content-related differences note the fact that television news has more emotionally laden content than does print media (Iyengar, 1991; McLuhan, 1964). Unlike print, television news has the ability to use "close-ups, zooms, slow motion, video graphics, and sound . . . [which] enables viewers to feel a 'presence' (Cho et al., 2003).

Consumer selectivity refers to the amount of freedom one has in choosing what news content to pay attention to. Selectivity among television viewers is reduced in comparison to those who read the newspaper. Reading the newspaper involves some amount of active seeking of information, whereas television news exposure is more passive. The amount of active seeking of information can be measured in part through asking how closely people pay attention to what they watch or read. Glancing at stories should have a different impact on emotions and cognitions than reading them.

A second area of interest in news types is the differences found between national news and local news. Several scholars note that many studies of the media-fear relationship are hindered by an exclusive focus on one media type, which restricts the ability to observe a possible differential impact of separate media types (Eschholz, 1997; Heath & Gilbert, 1996; Weitzer & Kubrin, 2004). In a study comparing six separate program types, including national and local news, Eschholz, Chiricos, & Gertz (2003) found that local news, but not national news, is associated with elevated crime fears.

More recently, Weitzer and Kubrin (2004) explored the relationship between fear of crime and frequency of exposure to several media types: radio, national news, local news, Internet, and newspaper. In a telephone survey of Washingtonians (N=480), the authors first estimate the ability of local news alone to predict fear, and compare this to the ability of the remaining media types (including local news) to predict fear. They find that those who identify local news as their principal news source are more fearful of crime than are those who report that national news or newspapers are their principal source of news.

Fear of Terrorism

A growing body of research explores the presence, extent, and consequences of fear of terrorism. Gallup Poll data on fear of terrorism have been collected and analyzed since 1995, originally developed in response to the bombing of the Alfred P. Murrah Federal Building in 1995 in Oklahoma City. The main terrorism-related question that has been used to measure fear of terrorism is: "How worried are you that you or someone in your family will become a victim of terrorism—very worried, somewhat worried, not too worried, or not worried at all?" The data show that just after the Oklahoma City bombings, 42% of Americans were "somewhat" or "very worried" that they or a family member could be a victim of a terrorist attack similar to the one in Oklahoma City (Jones, 2000). However, the last poll taken before September 11, 2001, showed that only 24% of Americans were very or somewhat worried (Saad, 2004). Research conducted on the day of the 2001 attacks indicates that 58% of the respondents were somewhat (35%) or very (23%) worried about being victimized by terrorism. Beginning in October 2001, the percentages began to decline and have fluctuated between 28% and 49% since.

Preliminary Gallup Poll data collected during the three months following September 11, 2001, show clear behavioral responses to the attacks: 40% of respondents handled their mail with more caution; 35% reported seeking information about bioterrorism; and 10% noted voluntary changes in their travel plans, avoidance of crowded places, stockpiling of goods, and/or purchasing of a weapon. Over 33% reported changing or planning to change their behavior as a result of September 11, 2001; this figure is more than two times the reported percentage following the Oklahoma City bombing.

In addition to public opinion polls such as these just mentioned, some research has examined terrorism reactions through studies devoted to the topic. Slone (2000) conducted an experimental study of the effects of media images on 237 viewers in a controlled environment. The experimental group was shown footage of Israeli news that depicted terrorist threats; the control group was shown footage of local Israeli news not consisting of terrorist threats. Slone's main hypothesis tested whether heightened anxiety levels would be experienced among those exposed to the intervention showing terrorist activities. Both groups were administered identical questionnaires following the viewing, and results were compared. These results demonstrated significantly higher anxiety levels among those exposed to terrorist-related news, providing some preliminary support for media effects on fear of terrorism.

Researchers from the University of Michigan collected public opinion data on public responses to the attacks shortly after September 11, 2001 (Traugott et al., 2002). This research measured the ways in which individuals responded to the terrorist attacks, with a specific focus on the resiliency of the American public to withstand and recover from the events. Two waves of interviews were conducted. The first occurred between September 17 and October 13, 2001, and the second was conducted between March 11 and April 16, 2002. A total of 764 persons were interviewed, 613 of whom were successfully recontacted for the second interview. Interviewees reported their sense of safety and security from future attacks, and the degree to which they paid attention to media reports of terrorism. A negative correlation between these two variables appeared in the first wave of interviews, but this relationship disappeared in the second wave of data collection, suggesting that effects were not long-lasting.

A study conducted by Rubin, Haridakis, Hullman, Sun, Chikombero, and Pornsakulvanich (2003) also concentrated on media effects on fear of terrorism. Researchers examined effects of frequency of television exposure on fear of terrorism and whether individual-level differences (i.e., locus of control, victimization experience, and gender) could explain the relationship between media consumption and fear of terrorism. Researchers administered surveys to 218 undergraduate students about six months after the terrorist attacks of September 11, 2001. Respondents reported their overall television exposure, their exposure to terrorism-related news stories, their motivation for watching these news stories, the perceived credibility that they attributed to what they viewed, the amount of attention they paid to what they viewed, and their crime victimization experience. Results indicated that the effect of overall television exposure and terrorism-related television exposure had no effect on fear of terrorism; however, motivation for viewing was significantly related to fear of terrorism. This suggests that those who intentionally watch terrorism-related news reports are more fearful and feel less safe from terrorist activity. In addition, crime victims were more fearful of terrorism.

Huddy, Feldman, Capelos, and Provost (2002) explored the possibility that personal and national risk of terrorism might be distinct constructs, and also might have differential impacts on behaviors and economic optimism. They present findings from telephone surveys of 1,221 residents of Long Island and Queens, NY, that were conducted in October–November 2001 in which respondents reported their perceptions of personal terrorism risk (which unfortunately also included a measure for perceived risk to a family member) and national terrorism risk. Principal components analysis results demonstrate that although correlated (0.60), personal and national perceived threats are conceptually distinct constructs that should be measured separately.

Multivariate analyses using personal threat and national threat as dependent variables revealed that gender was a highly significant predictor of personal and national perceived threat: females were substantially more fearful than were men. However, increased perceptions of personal threat were determined by education, party identification, race, and proximity to attacks, whereas these factors did not significantly predict national threat. This finding lends support to the argument that the subject of fear (e.g., oneself or a family member) produces different responses from global measures such as "fear of terrorism." Huddy and his colleagues also examined the impact of personal versus national threat perceptions on opinions about the national economy and also on behavior modifications. A heightened national threat is indicative of a skeptical outlook about the nation's economy as well as about one's own personal economic future, but elevated personal threat of terrorism does not appear to influence personal *or* national economic forecasting at all.

Finally, the authors find strong support for their hypothesis that personal threat influences precautionary behaviors, but only modest support for the impact of national threat perceptions on precautionary behaviors.

A final study discussed here was conducted by Nellis (2007), and it examined the influence of media exposure on fear of terrorism and perceived risk of terrorism. In addition, she measured the reported importance of fear and perceived risk in driving policy preferences.

Policy Preferences

It is important to consider attitudinal consequences of fear of terrorism. If policy support is driven by misinformed fears and perceptions of risk that are amplified by media messages, the public may support policies that it would not otherwise endorse. The final research concern here is the possible influence of worry and perceived risk on support for terrorism-related policy preferences. The next section first identifies a handful of studies exploring the relationship between crime reactions and crime policy support. This section also includes studies that examined the relationship between terrorism reactions and antiterrorism policy support.

Crime Policy Preferences

Extant research on the impact of fear on support for harsher crime policies has not produced consistent results. Some research shows a link between fear and more punitive attitudes (Baumer, Messner, & Rosenfeld, 2003; Keil & Vito, 1991; Langworthy & Whitehead, 1986; Sprott & Doob, 1997), whereas other studies fail to find a significant relationship (Baron & Hartnagel, 1996; Ouimet & Coyle, 1991), or, at the most, a very weak relationship (Longmire, 1996; Stinchcombe et al., 1980).

Langworthy and Whitehead (1986) examined the relationship between fear and public attitudes about crime policies. Relying on public opinion poll data from 2,100 adults nationwide, the authors tested fear of crime on punitiveness. They measured punitiveness by asking respondents whether they believed that the purpose of incarceration was rehabilitation (i.e., not punitive), or punishment (i.e., punitive). Fear of crime was an index of reported fears about several specific offenses. Path analysis results revealed a clear relationship between fear and liberalism on punitiveness, but subsequent probit analysis of fear and liberalism on punitiveness indicated a much more complex relationship. Overall, the authors suggest that punitive attitudes are better explained by scores on fear and liberalism than by standard demographic characteristics. Confidence in these results is hampered by the operationalization of the main variables, however.

Tyler and Weber (1982) tested competing theories about the origin of support for the death penalty using telephone survey data (N=200) of residents of a Chicago suburb. The instrumental perspective assumes that support for the death penalty is derived from fear or concern about crime, whereas the symbolic perspective holds that political-social attitudes drive policy support. Their empirical results lend support for the symbolic perspective, and they conclude that beliefs about the death penalty originate from political-social attitudes that remain constant over time. Unfortunately, their measure of death penalty support (an additive index of three items pertaining to death penalty support) is somewhat weak; it achieves an alpha coefficient of only 0.52. Moreover, the community that was studied was experiencing an unusual spike in crime, so results may be atypical.

Keil and Vito (1991) tested fear of crime as a predictor of support for the death penalty among 619 residents in Kentucky who participated in a telephone survey. The researchers found direct and indirect effects of fear on this measure of punitiveness. The more fear a subject reported (as measured through neighborhood safety), the more likely the subject was to support the death penalty. Age, race, and victimization experience operated indirectly through fear of crime to translate into greater odds of supporting the death penalty.

More recent studies suggest that punitive attitudes emerge from law and order policies that force crime issues (e.g., the proliferation of gangs and drugs) to the forefront, thus influencing fear and public opinion (Callanan, 2001). Yet, there is a lack of consensus on whether fear is a significant predictor of punitiveness. This could be due to poor operationalization of fear variables (Warr, 1994, 2000) and the punitiveness variables (Harris, 1986; Sims, 2003). Or, ambiguity about this relationship could be the consequence of using an incomplete model—one that does not allow for the deliberate promotion of fear to encourage more punitive attitudes about crime. Whatever the reason for the ambiguity of these results, they remain unclear and do not support the argument that punitiveness is a simple response to fear or perceived risk of victimization.

Antiterrorism Policy Preferences

Throughout history, American legislators have responded to threats against American security by strengthening mechanisms of formal social control. For instance, after the bombing of Pearl Harbor, federal legislation was enacted calling for the internment of Japanese Americans. The events of September 11, 2001, have elicited comparable responses, culminating in the passage of the USA PATRIOT Act (P.L. 107-56). It is well established that threats to national security prompt public support for harsher policies and restrictions on civil liberties (Marcus, Sullivan, Theiss-Moore, & Wood, 1995). Although support for such policies is initially quite strong, the duration of this support is usually short. For instance, in 2003, 66% of respondents to a national survey were somewhat or very concerned that measures to fight terrorism in the United States could result in restrictions on individual freedoms (Pew Research Center for People and the Press, 2003). This contrasts with national poll data collected just after September 11, 2001, in which more than 60% of respondents felt that the average person would need to give up some civil liberties in order to prevent future acts of terrorism (Greenberg, Craighill, & Greenberg, 2004).

Although pollsters routinely collect data on emotional, behavioral, and attitudinal reactions to terrorism and terrorism-related information, few comprehensive empirical studies have explored connections between the emotional and cognitive reactions (worry and perceived risk) to terrorism threats and support for antiterrorism policies. A review of the few studies that could contribute to an understanding of the relationship between fear, risk, and policy preferences is provided here.

Lerner, Gonzalez, Small, and Fischhoff (2003) measured the effects of different emotional responses to terrorism on policy preferences. The researchers based their work on Appraisal Tendency Theory, which holds that emotions arise from situations and also elicit specific cognitive appraisals. These emotions remain with the individual beyond the circumstance that caused them, transitioning into a filter through which subsequent experiences are passed. Lerner et al. (2003) examined the extent to which the emotions of fear and anger elicited preferences for antiterrorism policies. To accomplish this, they conducted two experiments in which participants selected from a national random sample first completed assessments of anxiety, acute stress, and vengeance.

Approximately six weeks later, respondents were questioned via WebTV.[1] The researchers' main findings were that anger and fear were significantly related to risk estimates and policy preferences. Subjects exposed to anger-inducing footage had lower estimates of risk and were supportive of more punitive policies. Subjects exposed to fear-inducing footage had higher estimates of risk and were more supportive of preventive policies. Although rigorous in methodology, this study used an experimental design that poses threats to external validity that often accompany research conducted in artificial environments.

Willer (2004) tested Social Identity Theory, which claims that when there is fear of external attack, support for standing leaders will increase. Willer hypothesized that issuances of terror warnings by the federal government would lead to increased support for the president. Using secondary data collected by the Gallup Organization and the *Washington Post*, he lagged the effects of reported changes in the level of threat of terrorist activity on presidential approval ratings and controlled for important possible intervening variables, including the war in Afghanistan, the war in Iraq, Saddam Hussein's capture, and photos released and published from the Abu Ghraib prison. Willer examined lagged effects of changes in the threat of terrorist activity on public perceptions of the economy as a second dependent variable. Twenty-six government-issued reports of possible terrorist attacks reported in the newspaper were recorded from February 1, 2001, through May 9, 2004. Gallup Poll data on the presidential general and *economic* approval ratings were also collected for roughly this period. Results of four separate regression models for each of the dependent variables indicate a statistically significant relationship between these two measures, suggesting that the public becomes more supportive of the president following government-issued terror warnings. The duration of these effects was not determined, however.

Huddy, Feldman, Taber, and Lahav (2005) tested two hypotheses in their measurement of the influence of anxiety and perceived threat on policy preferences that are relevant to the present study. First, they expected that elevated perceptions of threat would result in stronger support for militaristic responses to the terrorist attacks on September 11,

1. WebTV enables people to access the Internet through their televisions on a limited basis.

2001, whereas heightened anxiety would not because fearful individuals are risk averse. Second, they expected that perceived threat of terrorism would be associated with more support for domestic antiterrorism policies because these presumably lower future risks. Anxiety would also lead to more support for such policies because the policies' hopes for the ability to reduce risk, which, in turn, alleviates fear.

The authors tested these arguments empirically with national telephone survey data of 1,549 adults. Results lend support for their first hypothesis: anxiety and threat have a conflicting influence on support for retaliatory military action in the expected directions—threat is associated with more support, and fear is associated with less support. As regards domestic antiterrorism policies (operationalized as support for a national I.D. card and support for surveillance of telephone and e-mail communications), the authors observed that increased perceptions of threat were associated with greater support for antiterrorism policies, but they did *not* observe a significant effect of fear on these policies.

Greenberg, Craighill, and Greenberg (2004) analyzed two sets of data collected by Pew Research Center in 2001 and 2002 that asked several questions about antiterrorism measures, including support for allowing the government to monitor credit card purchases, telephone calls, and e-mails, and supporting a governmental requirement to carry a national I.D. card. The authors were interested to know how support for these policies changed between 2001 and 2002, and which emotional reactions to terrorism motivated a willingness to relinquish some civil rights. Their results indicate that support for civil liberty infringements lessens over time, though differences in question wording across the datasets renders these findings somewhat untrustworthy. Also, several emotional reactions were observed as closely associated with support for these antiterrorism measures, including fear, anger, depression, and insomnia.

One problematic feature of this study is that the second round of data was collected during August 2002, which was around the time of the one-year anniversary of September 11, 2001. Although this was done deliberately so that researchers could measure the differences in willingness to give up some civil liberties one year after the attacks, it nevertheless is assumed that the first anniversary of this catastrophic event had some impact on responses.

A national survey of 1,448 adults explored one's willingness to exchange civil liberties for a greater sense of personal security (Davis & Silver, 2004). The authors expected to find that trust in government and fear would interact to produce a greater willingness to concede on select antiterrorism policy preferences. Their results support this hypothesis—a greater sense of threat combines with a greater trust in government to produce an increased willingness to relax some civil liberties.

RECENT RESEARCH: BUILDING ON PREVIOUS FINDINGS

In 2006, 532 residents of New York City and Washington, DC, were interviewed on the topic of responses to terrorism-related information (Nellis, 2007). This study investigated the effects of media exposure on fear of terrorism, perceived risk of terrorism, and support for antiterrorism policy preferences among residents in the two cities. The aim of the study was to determine whether a consequence of fear or heightened risk of terrorism would translate to greater support for harsher antiterrorism measures. If fear and perceived risk lead to support for policies that might not otherwise be supported, this is important to know, especially if fears are elevated by media "hype."

Findings suggest that individuals are still fearful of terrorism and perceive the *risk* of terrorism to be high. Moreover, respondents who watch greater amounts of television news report higher levels of fear and perceived risk; in some instances, these effects are amplified by the amount of attention paid to terrorism-related news and the credibility they assign to terrorism-related news. Victims report greater terrorism fear and higher terrorism risk than do nonvictims, and this research finds some evidence that frequency of news exposure alters their fears and risk perceptions. Finally, results show that news effects differ by media type. Specifically, respondents report greater overall news effects when controlling for television news in comparison to news obtained from the newspaper.

This study also examined the impact of fear and perceived risk on support for antiterrorism policy preferences. Although it was expected that greater terrorism fear and higher risk perceptions would be associated with enhanced support for more stringent civil liberties regulations, and especially punitive responses to suspected and convicted terrorists, this was not consistently the case.

This study tested 10 separate logistic regression models: five models tested the impact of fear of more terrorist attacks on the likelihood of supporting five separate antiterrorism policies, and five models tested the impact of perceived risk of an attack in the next 12 months on the likelihood of supporting these same policies. There is support for this possibility here, though it could be the case that those who are more fearful watch more news *and* endorse stricter policies. Fear is a significant motivator for support of three of the five antiterrorism policies examined, whereas perceived risk of an attack in the next 12 months is associated with only one antiterrorism policy preference: requiring citizens to carry a national I.D. at all times. Furthermore, for two antiterrorism policies (supporting personal communications monitoring and supporting the death penalty for convicted terrorists), fear is associated with higher likelihood of supporting the policy, whereas perceived risk is not. This corroborates findings elsewhere that fear and risk are different constructs, but also that fear may be even stronger than risk in regard to antiterrorism policy support.

CONCLUSIONS

Since September 11, 2001, newsmakers have inundated audiences with frightening images, threats, and stories on the topic of terrorism. Over the past five years, the public has learned of countless false terrorist alarms and thwarted terrorist plots, and has been asked to participate in defending the homeland through vigilance of suspicious activity or persons. The country's government-issued Terror Alert System will remain at the "elevated" level for the foreseeable future, indicating a significant risk of terrorism. In New York City, this alert is at the "high" level, translating into a high risk of terrorist attacks. I expect that the topic of terrorism will continue to dominate the news for some time.

Important questions that current social science research attempts to answer include the following: How does the public cope with terrorism-related information? Specifically, are residents afraid of terrorism? Do they perceive the risk of terrorism to be high? What relationships exist among fear, risk, and media exposure? Are emotional and cognitive responses to terrorism similar across media types? And, do the factors of frequency of news exposure, attention paid to terrorism news, and credibility of terrorism news amplify or mitigate fears and risk perceptions? Similarly, what differences in risk do individuals perceive for themselves, for others, and for terrorist events in general? Finally, to what extent are fears and risk perceptions associated with various antiterrorism policy preferences?

Nellis (2007) and others find that frequency of exposure to the news is a strong predictor of terrorism fear and risk. However, the cross-sectional data structure in most studies limits a complete understanding of the order of the news–fear relationship: it is possible that one way in which fearful individuals cope with terrorism-related information is to follow the news more frequently, or it could be that the news is scaring people and leading them to calculate a disproportionately high terrorism risk.

Excessive fear of terrorism and inaccurate terrorism risks can have a number of detrimental effects, one of which is support for punitive antiterrorism policies. For this reason, it is important to attempt to identify the factors that contribute to fear and risk. If the media is merely conveying an accurate depiction of terrorist activity, and if individuals are using rational calculation to estimate risk of terrorist attacks, then the positive relationship between television news exposure and perceived risk and fear is an understandable outcome. If, however, the media is sensationalizing or exaggerating terrorism stories, this may be heightening the perceptions of risk and fear in the population as a whole.

There are serious policy implications that could stem from this: if objective conditions are inflated by media hype and this leads to elevated fears among viewers, there exists a risk of endorsing support for overly punitive policies and restrictions on civil liberties that might not otherwise be supported.

REFERENCES

Altheide, D. (2002). *Creating fear: News and the construction of crisis.* New York: Wafter de Gruyter.

Altheide, D. (2004). "Consuming terrorism." *Symbolic interaction* 27(3): 289–308.

Austin, D. M., Furr, L. A., & Spine, M. (2002). "The effects of neighborhood conditions on perceptions of safety." *Journal of criminal justice* 30: 417–427.

Baron, S. & Hartnagel, T. F. (1996). "'Lock 'em up:' Attitudes toward punishing juvenile offenders." *Canadian journal of criminology and criminal justice* 38(2): 191–212.

Baumer, E. P., Messner, S. F., & Rosenfeld, R. (2003). "Explaining spatial variation in support for capital punishment: A multilevel analysis." *American journal of sociology* 108(4): 844–875.

Bureau of Transportation Statistics. (2007). *Transtats Homepage*. Accessed August 24, 2007, from http://www.transtats.bts.gov.

Callanan, J. (2001). *The determinants of punitiveness: The effects of crime related media consumption and crime salience on California's support for three strikes sentencing*. Doctoral dissertation, University of California–Irvine.

(2002). "The racial and ethnic typification of crime and the criminal typification of race and ethnicity in local television news." *Journal of research in crime and delinquency* 39:400–420.

Chiricos, T., McEntire, R., & Gertz, M. (2001). "Perceived racial and ethnic composition of neighborhood and perceived risk of crime." *Social problems* 48(3): 322–340.

Cho, J., Boyle, M. P., Keum, H., Shevy, M. D., McLeod, D. M., Shah, D. V., et al. (2003). "Media, terrorism, and emotionality: Emotional differences in media content and public reactions to the September 11th terrorist attacks." *Journal of broadcasting and electronic media* 47(3): 309–327.

Clemente, F. & Kleiman, M. B. (1979). "Fear of crime among the aged." *Gerontologist* 16(3): 207–210.

Covington, J., & Taylor, R. (1991). "Fear of crime in residential neighborhoods: Implications of between- and within-neighborhood sources for current models." *Sociological quarterly* 32: 321–349.

Davis, D. W. & Silver, B. D. (2004). "Civil liberties vs. security: Public opinion in the context of the terrorist attacks on America." *American journal of political science* 48(1): 28–46.

Eschholz, S. (1997). "The media and fear of crime: A survey of the research." *Journal of law and public policy* 9(1): 37–59.

Eschholz, S., Chiricos, T., & Gertz, M. (2003). "Television and fear of crime: Program types, audience traits, and the mediating effect of perceived neighborhood racial composition." *Social problems* 50(3): 395–415.

Farrall, S., Bannister, J., Ditton, J., & Gilchrist, E. (1997). "Questioning the measurement of 'fear of crime.'" *British journal of criminology* 37(4): 658–679.

Feeley, M. & Simon, J. (1994). "Actuarial justice: The emerging new criminal law." In D. Nelkin (Ed.), *The future of criminology* (p. 172–201). Thousand Oaks, CA: Sage.

Ferraro, K. (1995). *Fear of crime: Interpreting victimization risk*. Albany: State University of New York Press.

Ferraro, K. & LaGrange, R. L. (1992). "Are older people most afraid of crime? Reconsidering age differences in fear of victimization." *Journal of gerontology* 47(5): S233–S244.

Fisher, B. S. & Sloan, J. J. (2003). "Unraveling the fear of victimization among college women: Is the 'Shadow of Sexual Assault' hypothesis supported?" *Justice quarterly* 20(3): 633–659.

Flynn, J., Slovic, P., & Mertz, C. K. (1993). "Decidedly different: Expert and public views of risks from a radioactive waste repository." *Risk analysis* 13(6): 643–648.

Garofalo, J. & Laub, J. (1978). "The fear of crime: broadening our perspective." *Victimology: An international journal* 3(3/4): 242–253.

Greenberg, M., Craighill, P., & Greenberg, A. (2004). "Trying to understand behavioral responses to terrorism: Personal civil liberties, environmental hazards, and U.S. resident reactions to the September 11, 2001 attacks." *Human ecology review* 11(2): 165–176.

Hale, C. (1996). "Fear of crime: A review of the literature." *International review of victimology* 4: 79–150.

Harris, P. (1986). "Oversimplification and error in public opinion: Surveys on capital punishment." *Justice quarterly* 3(4): 429–456.

Heath, L. & Gilbert, K. (1996). "Mass media and fear of crime." *American behavioral scientist* 39: 379–386.

Huddy, L., Feldman, S., Capelos, T., & Provost, C. (2002). "The consequences of terrorism: Disentangling the effects of personal and national threat." *Political psychology* 23(3): 485–509.

Huddy, L., Feldman, S., Taber, C., & Lahav, G. (2005). "Threat, anxiety, and support of antiterrorism policies." *American journal of political science* 49(3): 593–608.

Iyengar, S. (1991). *Is anyone responsible? How television frames political issues.* Chicago: University of Chicago Press.

Jackson, J. (2006). "Introducing fear of crime to risk research." *Risk analysis* 26(1): 253–264.

Jones, J. (2000). *Americans less concerned about terrorist attacks five years after Oklahoma City*. Washington, DC: Gallup Organization.

Kanan, J. W. & Pruitt, M. V. (2002). "Modeling fear of crime and perceived victimization risk: The (in)significance of neighborhood integration." *Sociological inquiry* 72(4): 527–548.

Kasperson, J. X., Kasperson, R. E., Pidgeon, N., & Slovic, P. (2003). "The social amplification of risk: Assessing fifteen years of research and theory." In N. Pidgeon, R. E. Kasperson, & P. Slovic (Eds.), *The social amplification of risk* (p. 13–46). Cambridge, UK: Cambridge University Press.

Kasperson, R. E., Renn, O., Slovic, P., Brown, H. S., Emel, J., Goble, R., et al. (1988). "The social amplification of risk: A conceptual framework." *Risk analysis* 8(2): 177–187.

Keil, T. J. & Vito, G. F. (1991). "Fear of crime and attitudes toward capital punishment: A structural equations model." *Justice quarterly* 8(4): 447–464.

Kelling, G. L. & Coles, C. M. (1996). *Fixing broken windows.* New York: Simon & Schuster.

Kubrin, C. E. & Weitzer, R. (2003). "New directions in social disorganization theory." *Journal of research in crime and delinquency* 40(4): 374–402.

LaGrange, R. L., Ferraro, K. F., & Supancic, M. (1992). "Perceived risk and fear of crime: The role of social and physical incivilities." *Journal of research in crime and delinquency* 29(3): 311–334.

Lane, J. & Meeker, J. W. (2000). "Subcultural diversity and fear of crime and gangs." *Crime and delinquency* 46(4): 497–521.

Lane, J. & Meeker, J. W. (2003). "Women's and men's fear of gang crimes: Sexual and nonsexual assault as perceptually contemporaneous offenses." *Justice quarterly* 20(2): 337–371.

Langworthy, R. H. & Whitehead, J. T. (1986). "Liberalism and fear as explanations of punitiveness." *Criminology* 24(3): 575–591.

Lee, M. S. & Ulmer, J. T. (2000). "Fear of crime among Korean Americans in Chicago communities." *Criminology* 38(4): 1173–1206.

Lerner, J. S., Gonzalez, R. M., Small, D. A., & Fischhoff, B. (2003). "Effects of fear and anger on perceived risks of terrorism: A national field experiment." *Psychological science* 14(2): 144–150.

Lewis, D. A. & Maxfield, M. G. (1980). "Fear in the neighborhoods: An investigation of the impact on crime." *Journal of research in crime and delinquency* 17(2): 160–189.

Lewis, D. A. & Salem, G. (1986). *Fear of crime: Incivility and the production of a social problem.* New Brunswick, NJ: Transaction.

Longmire, D. (1996). "Americans' attitudes about the ultimate weapon: Capital punishment." In T. J. Flanagan & D. R. Longmire (Eds.), *Americans view crime and justice* (p. 93–108). Thousand Oaks, CA: Sage.

Marcus, G. E., Sullivan, J. L., Theiss-Moore, E. Y., & Wood, S. L. (1995). *With malice toward some: How people make civil liberties judgments.* Cambridge, UK: Cambridge University Press.

Madriz, E. I. (1997). "Images of criminals and victims: A study on women's fear and social control." *Gender and society* 11(3): 342–356.

Markowitz, F. E., Bellair, P. E., Liska, A. E., & Liu, J. (2001). "Extending social disorganization theory: Modeling the relationships between cohesion, disorder, and fear." *Criminology* 39(2): 293–321.

McLuhan, M. (1964). *Understanding media: The extensions of man.* Cambridge, MA: MIT Press.

Menard, S. (2002). *Applied logistic regression analysis.* Thousand Oaks, CA: Sage.

Nellis, A. (2007). *How does the American public cope with terrorism-related information?* Doctoral dissertation, American University.

Ouimet, M. & Coyle, E. J. (1991). "Fear of crime and sentencing punitiveness: Comparing the general public and court practitioners." *Canadian journal of criminology and criminal justice* 33(2): 149–162.

Pew Research Center for People and the Press. (2003). *Views of a changing world.* Washington, DC: Author.

Romer, D., Jamieson, K. H., & Aday, S. (2003). "Television news and the cultivation of fear of crime." *Journal of communication* 53(1) 88–104.

Rountree, P. W. & Land, K. C. (1996). "Perceived risk versus fear of crime: Empirical evidence of conceptually distinct reactions in survey data." *Social forces* 74(4): 1353–1376.

Rubin, A. M., Haridakis, P. M., Hullman, G. A., Sun, S., Chikombero, P. M., & Pornsakulvanich, V. (2003). "Television exposure not predictive of terrorism fear." *Newspaper research journal* 24(1): 128–145.

Saad, L. (2004, September 10). *Three years after 9/11, most Americans carrying on normally.* Washington, DC: Gallup Organization.

Sacco, V. F. (1995). "Media constructions of crime." *Annals of the American academy of political and social science* 539: 141–154.

Shaw, C. R. & McKay, H. D. (1969). *Juvenile delinquency and urban areas* (Rev. ed.). Chicago, IL: University of Chicago Press.

Sims, B. (2003). "The impact of causal attribution on correctional ideology: A national study." *Criminal justice review* 28(1): 1–25.

Skogan, W. G. (1990). *Disorder and decline.* New York: Free Press.

Skogan, W. G. (1995). "Crime and racial fears of white Americans." *Annals of the American academy of political and social science* 539: 59–71.

Skogan, W. G. & Maxfield, M. G. (1981). *Coping with crime: Individual and neighborhood reactions.* Beverly Hills, CA: Sage.

Slone, M. (2000). "Responses to media coverage of terrorism." *Journal of conflict resolution* 44(4): 508–522.

Slovic, P. (1987). "Perception of risk." *Science* 236: 280–285.

Slovic, P. (2000). "Perceived risk, trust, and democracy." In P. Slovic (Ed.), *The perception of risk* (p. 316–326). London: Earthscan.

Slovic, P., Fischhoff, B., & Lichtenstein, S. (1979). "Rating the risks." *Environment* 21(3): 14–20, 36–39.

Slovic, P. & Weber, E. U. (2003). *Perception of risk posed by extreme events.* Working paper, Columbia University.

Sprott, J. B. & Doob, A. (1997). "Fear, victimization, and attitudes to sentencing, the courts, and the police." *Canadian journal of criminology and criminal justice* 39(3): 275–291.

Stinchcombe, A. L., Adams, R., Heimer, C. A., Scheppelle, K. L., Smith, T. W., & Taylor, D. G. (1980). *Crime and punishment: Changing attitudes in America.* San Francisco: Jossey Bass.

Taylor, R. (2001). *Breaking away from broken windows.* Boulder, CO: Westview Press.

Taylor, R. (2002). "Fear of crime, social ties, and collective efficacy: Maybe masquerading measurement, maybe déjà vu all over again." *Justice quarterly* 19(4): 773–792.

Traugott, M., Brader, T., Coral, D., Curtin, R., Featherman, D., Groves, R., et al. (2002, September). *How Americans responded: A study of public reactions to 9/11/01*. Institute for Social Research, University of Michigan.

Tulloch, M. (2000). "The meaning of age differences in the fear of crime." *British journal of criminology* 40: 451–467.

Tyler, T. R. (1980). "Impact of directly and indirectly experienced events: The origin of crime-related judgments and behaviors." *Journal of personality and social psychology* 39(1): 13–28.

Tyler, T. R. & Weber, R. (1982). "Support for death penalty: Instrumental response to crime or symbolic attitude?" *Law and society review* 17(1): 21–45.

Warr, M. (1980). "The accuracy of public beliefs about crime." *Social forces* 59(2): 456–470.

Warr, M. (1984). "Fear of victimization: Why are women and the elderly more afraid?" *Social science quarterly* 65(3): 681–702.

Warr, M. (1994). "Public perceptions and reactions to violent offending victimization." In A.J. Reiss and J. A. Roth (Eds.), *Understanding and preventing violence: Consequences and control* (Vol. 4, p. 1–66). Washington, DC: National Academic Press.

Warr, M. (2000). *Fear of crime in the United States: Avenues for research and policy.* Washington, DC: National Institute of Justice.

Warr, M. & Stafford, M. (1983). "Fear of victimization: A look at the proximate causes." *Social forces* 61(4): 1033–1043.

Weinstein, N. D. (1989). "Optimistic biases about personal risks." *Science* 246(4935): 1232–1233.

Weitzer, R. & Kubrin, C. E. (2004). "Breaking news: How local TV news and real-world conditions affect fear of crime." *Justice quarterly* 21(3): 497–520.

Willer, R. (2004). "The effects of government-issued terror warnings on presidential approval ratings." *Current research in social psychology* 10(1): 1–12.

Wilson, J. Q. & Kelling, G. L. (1982, March). "Broken Windows." *Atlantic Monthly* p. 29–38.

CHAPTER 7

Conclusion

Robin Valeri and Kevin Borgeson

Terrorism, as demonstrated by the various topics and approaches presented here, is a multifaceted issue. Therefore, attempting to understand terrorism will necessarily be complex. As it is currently defined, the study of terrorism includes attempting to understand the causes of terrorism. What prompts an individual or group of individuals to engage in terrorist activities? What environmental factors play a role in shaping someone to become a terrorist? It also includes efforts to combat terrorism as well as efforts to understand why certain counterterrorism policies and practices are implemented, whereas others—perhaps ones that would be more effective—are not implemented. As was discussed, the United States government, through the establishment of agencies, laws, and policies, has tried to combat terrorism. Additionally, law enforcement agencies from the local to the federal level are working to combat terrorism. Finally, there are the consequences of terrorism. These include dealing with the aftermath of a terrorist act; providing support and counseling to victims, their families, and all of the people involved in search, rescue and recovery; and coping with the resulting fear of terrorism.

O'Connor presents several theories of terrorism. Most relevant to understanding the beliefs and actions of white supremacist hate groups in the United States and their interactions with Muslim extremists are the Political Theory of Fascism and Religion as a Theory of Terrorism. As O'Connor discusses, "fascism appeals to the frustrations and resentments of a race of people who think they ought to have a bigger place at the global table." As is made clear by Borgeson and Valeri in their discussion of Christian Identity, members of the Aryan Nations believe that whites are the true inheritors of God's dominion. Jews are portrayed as the offspring of the devil. Additionally, members of the Aryan Nations believe that the United States rightfully belongs to white Americans. According to the Aryan Nations, the Jews have not only been successful at gaining control of the media, politics, and banking, but have convinced the liberal population that they, the Jews, are the true inheritors of God's Promised Land. Thus, consistent with fascism as a theory of terrorism, members of the Aryan Nations perceive the Jews as a threat to what they, white Americans, are rightfully entitled. They believe that the Jews are encroaching upon and taking over their country, and they believe that the Jews are the source of past and present problems. Both fascist theories of terrorism and religious theories of terrorism view the enemy as a demon, and for this reason see the actions of their own group against the enemy as not only justifiable, but sanctioned by God. Again, the doctrine of Christian Identity includes the belief that the Jews are the offspring of the devil. As O'Connor notes, fascist theories of terrorism support terrorism at home and abroad. As is discussed by Borgeson and Valeri, ". . . our [Aryan Nations] Jihad is . . . for the extermination of the satanic 'Jew,' worldwide, we see no borders, nor boundaries." Thus, the goal of the Aryan Nations is to begin by taking back the United States, ridding it of Jews and other nonwhites, and then to rid the rest of the world of Jews. It is sharing a common enemy, the Jew, with Islamic Jihadists that has prompted some Aryan Nations members to consider working with Islamic Jihadists to accomplish their shared goal. As O'Connor points out, theological theories of terrorism typically see the enemy as a foreign influence. O'Connor also discusses the importance of martyrs to theology-based terrorism. The use of suicide bombers accomplishes the dual goals of creating an act of terrorism and creating a martyr.

O'Connor also provides a brief coverage of sociological and psychological theories of terrorism. However, he fails to cover basic social psychological explanations of terrorism. These explanations stem from theories and research on groups and attitudes, especially prejudice, hate, terror management, and social identity. Borgeson and Valeri, in their discussion of attempts by the Aryan Nations to establish ties with Islamic extremists, use theories of prejudice by Allport, Blumer, and Sherif as well as

Sternberg's theory of hate to provide the framework for understanding the actions of hate groups.

Terror management theory (Greenberg, Solomon, & Pyszczynski, 1997; Solomon, Greenberg, & Pyszczynski, 1991), which attempts to explain why people need self-esteem, also can be used as a framework for understanding terrorism. According to terror management theory, humans, like other animals, have an instinctive motivation for self-preservation. However, as humans, we are aware of and able to reflect on the fact that our own death is inevitable. Our desire for self-preservation conflicts with the knowledge that we will die. Thinking about our own death creates anxiety or terror. According to terror management theory, we manage the resulting terror by adopting our culture's perspectives about life, the meaning of life, and death transcendence, whether it is through reincarnation, an afterlife, or some other means. Thus, an individual's culture is instrumental in shaping self-esteem because one's culture determines what behaviors and beliefs are valued and provides a prescription for the behaviors necessary to transcend death. In order to transcend death, an individual must come to see himself or herself as a valuable and significant member of society and thus eligible for immortality. In sum, having a high self-esteem protects an individual from mortality concerns. To this end, there is ample evidence examining the relationship between self-esteem and mortality salience. Research (Greenberg et al., 1992) suggests that heightened self-esteem provides a buffer against anxiety resulting from thoughts of one's own death. Other research suggests that when mortality salience is increased, efforts to bolster self-esteem increase (Pyszczynski et al., 2004).

Pyszczynski, Solomon, and Greenberg (2002), in their book *In the Wake of 9/11: The Psychology of Terror*, use terror management theory as a means for understanding the impact of 9/11 as well as for understanding the roots of terrorism. Although their book focuses on understanding Islamic terrorism, terror management theory can be used to explain terrorism in general and can be applied to domestic terrorism. The following quote explains the threat posed by different others:

> The mere existence of those who are different poses a threat to the individual's faith in the absolute validity and correctness of his or her own perspective on reality. This threat undermines the protections against deep existential fears that our worldviews provide. Derogating different

> others, working to convert them to one's own worldview, and in some cases doing battle to eliminate them from the face of the earth and thus help create a more perfect world can thus become a heroic, self-esteem-enhancing, and salvation-assuring virtuous activity that adds further protection to the defensive shield that we all need to survive in a world where the only real certainty is our inevitable demise. (Pyszczynski, Solomon, & Greenberg, 2002, p. 153)

According to Pyszczyski, Solomon, and Greenberg (2002), cultures that promote tolerance and value diversity are less likely to strike out against others even when mortality salience is high. In contrast, cultures that promote intolerance, as well as people who are high in authoritarianism or prone to low self-esteem, are more likely to aggress against different others, especially when mortality salience is high. Certainly, white supremacist groups and followers of Christian Identity are not noted for their tolerance. Additionally, leadership among these groups tends to be authoritarian. These characteristics begin to set the stage for terrorist activities. Prejudice is more likely to turn to violence when there is someone or some group to strike out against. For white supremacist groups and Christian Identity adherents, Jews—and institutions such as the government and media that are believed to be controlled by Jews—serve as useful scapegoats. According to terror management theory, these groups threaten the existential protection that one's culture provides. This is especially true when those espousing different values and living according to different principles appear to be successful and prosperous.

Pyszczyski, Solomon, and Greenberg (2002) further explain that many of the people who commit or are likely to commit terrorism not only have belief systems or personalities that support intolerance, but also live in conditions that result in heightened mortality salience. These conditions include violence, war, famine, and disease. Although these conditions may not seem to be applicable to domestic terrorists, it is important to understand that white supremacists believe they are in a war against evil and are fighting an enemy, the Jew, who is successfully encroaching upon and taking over what is rightfully theirs.

Similar to terror management theory, cultural values, personal identity, and self-esteem play important roles in social identity theory. According to social identity theory (Tajfel & Turner, 1979; 1986), our self-image is composed of our own personal identity as well as our social identity. A person's social identity derives from the individual's various group memberships. These include ethnicity, gender, and religion as well as other groups to which the individual belongs or joins. Tajfel and Turner (1979; 1986) suggest that individuals—because they are motivated to evaluate themselves positively—also are motivated to evaluate the groups to which they belong positively. Essential to the evaluation process is a comparison between groups. According to social identity theory, this comparison will be biased in such a manner that the in-groups compare favorably to the out-groups.

Taylor (1997, 2002) departed from traditional views of the self by proposing a model of the self-concept in which collective identity has psychological priority over both collective esteem and personal identity, which in turn has priority over personal esteem. According to Taylor, developing a personal identity first requires that an individual determine his or her unique characteristics. To do this, the individual must compare his or her attributes to norms or standards that are derived from the individual's own group. Thus, an individual's social identity—or, in Taylor's terminology, an individual's collective identity—provides the basis for comparison. Thus, personal identity, or the establishment of one's unique attributes, derives from one's collective identity, and personal or self-esteem is an evaluation of one's personal identity, just as collective esteem is an evaluation of one's collective identity.

Taylor and Louis (2005) propose that members of terrorist groups look to their collective identity not only to determine the group's beliefs, values, and behaviors, but also as a basis for establishing their own personal identity. Whereas Social Identity Theory posits that our social identity derives from our various group memberships, Taylor and Louis propose that some groups take precedence over others. Specifically, an individual's cultural—or, in some cases, religious—group, because of its pervasiveness throughout the individual's life, takes precedence over other groups in shaping identity. For an individual whose life is centered on their religion, their social identity as a member of that religious group not only provides them with religious beliefs, but also with a shared history, valued goals, and the steps for achieving those goals. Taylor and Louis further explain that for terrorists, their collective identity and the norms of their in-group are constantly salient because of their minority status within their own society as well as their chronic orientation toward the conflict with the out-group. As a result, in-group norms play an especially powerful role in shaping the terrorist's beliefs and behaviors.

According to Taylor and Louis (2005), the out-group also plays an important role in defining the identity of the terrorists. If the terrorist group is able to identify an evil enemy as the source of their suffering, than they can blame this group for any disadvantages or suffering they have endured. Taylor and Louis further suggest that terrorist groups, when faced with a low probability of achieving positive outcomes for their own group, may come to view achieving negative outcomes for the enemy as a benefit to their own group. Thus, terrorist acts, although they may not directly benefit one's own group, may be seen as an end in themselves.

Somewhat related to this is research by Struch and Schwartz (1989) that examined the attitudes of Israeli citizens toward an ultraorthodox religious sect. In addition to standard measures of in-group bias, these researchers also included measures of aggression. Results revealed positive correlations between perceived conflict of interests and aggression as well as between intergroup value dissimilarity and aggression. Thus, the more one group perceives its interests as conflicting with the other group's interests—or the more one group perceives its values as diverging from the other group's values—the more likely that group is to aggress against the other. Finally, the positive correlation between perceived conflict of interest and aggression was strongest for those groups of Israelis who themselves identified strongly with some religious (but not ultraorthodox) group. These findings are important to understanding the actions of domestic terrorists because the majority of domestic terrorist groups are right-wing groups associated with the religious ideology Christian Identity. Thus, any of these groups who have strong religious beliefs may be more predisposed to engage in terrorist activities.

Taken together, these social psychological theories for understanding terrorism—although they each offer a unique perspective, when broken down into their component parts—share the perspective that terrorism begins with differences between two groups. These differences are exacerbated by some threat, whether it is a threat to one's immortality, values, welfare, or security. The out-group is blamed for both past and present injustices. As part of this process, the out-group is denigrated or demonized. Thus, the behaviors and beliefs of the out-group that make them different become the cornerstone of what makes them less than equal to the in-group—and, as a consequence, less worthy of equal treatment, protection, and justice. Striking out against the out-group is legitimized by the beliefs of the in-group. "They" are different from us, "They" are less than us, "They" are taking from us, and, therefore, "They" deserve to be destroyed. This "us versus them" dichotomy and demonization of the out-group is not only a key component to social psychological theories of terrorism, but also is a key component to many of the theories of terrorism discussed by O'Connor. This "us versus them" scenario and demonization of the enemy is clearly evident in the presentation of Christian Identity and in the description of Aryan Nations members.

Although these theories of terrorism help us gain a better understanding of terrorism, can they also help us to develop a means for minimizing the threat of terrorism? The focus of counterterrorism is uncovering terrorist plots. As is discussed by Costanza, Kilburn, and Helms in Chapter 5, counterterrorism measures in the United States have typically focused on uncovering threats from the left, such as threats arising from labor organizations, communists, environmental groups, or animal rights activists. Interestingly, Costanza, Kilburn, and Helms identify the Palmer Raids as the first systematic attempt at counterterrorism within the United States. These raids targeted immigrants and frequently resulted in their deportation without trial. The public's reaction to these raids was largely positive. Similarly, after the first bombing of the World Trade Center in 1993 and the bombing of the Oklahoma City Federal Building in 1995, a series of laws to combat undocumented immigration, crime, and terrorism came into effect. As pointed out by Costanza and colleagues, these laws were largely supported by the public and focused on foreign sources of terrorism. Given that, prior to 9/11, the bombing of the Oklahoma City Federal Building was the deadliest terrorist attack on American soil and was perpetrated by an American citizen, it seems that these measures, although well received by the public, failed to address a key source of terrorism within the United States—specifically, domestic terrorist groups. After 9/11, there was a resurgence of interest in immigrants and foreigners as the source of terrorism. For example, in the weeks following 9/11, the USA PATRIOT Act was passed; this was followed by the Homeland Security Act of 2002. Whereas the post-9/11 enactment of antiterrorism laws focused on foreigners is understandable, taken together with the antiterrorism laws passed in the 1990s, which also focused on foreigners, two questions arise: What is being done to combat domestic terrorism? And, why does the public seem content to ignore domestic sources of terrorism?

Nellis discusses the role that fear plays in responding to terrorism. As Nellis points out, the policy decisions made by a fearful public may be quite different from those made by a rational, well-informed public. The public's fear stems in part from a perceived lack of control. Although an individual may believe that they can take reasonable steps to minimize their chances of becoming a victim of crime, most people feel that there is little they can do to minimize the threat of terrorism. For example, an individual—to minimize their chance of becoming a victim—may install a security system in their house or avoid unsafe neighborhoods. However, these same or similar actions do not serve as a deterrent against terrorism. Research in social psychology (Langer & Rodin, 1976; Rodin, 1986; Rodin & Langer, 1977; Schultz, 1976; Taylor et al., 2000) suggests that the perception of control enhances our ability to effectively cope with stress. We react negatively to a loss of control (Brehm & Brehm, 1981; Taylor, Lichtman, & Wood, 1984; Thompson et al., 1993; Twenge, Zhang, & Im, 2004), and, in situations beyond our control, we may work to create the illusion of control (Biner et al., 1995; Thompson, 1999). For example, a student may sleep with his pajamas on inside-out in the belief that this behavior will make a snow day tomorrow more likely. Unfortunately, the illusion of control, when it is just that, an illusion, can be maladaptive (Affleck et al., 1987; Colvin & Block, 1994; Helgeson, 1992). In the case of terrorism, policies may be implemented because they create an illusion of control. However, they may be maladaptive, as Nellis points out, because they are overly punitive or restrict civil liberties. Or, they may be maladaptive because they result in too many false positives, falsely identifying someone as a terrorist, and too many false negatives (incorrectly labeling a real threat as harmless).

The media also plays an important role in shaping the public's beliefs about terrorism and the perceived risk we face. The public's judgment regarding the likelihood and severity of terrorist threats is shaped largely by the news, which in turn impacts the decisions made by politicians. As Nellis points out, our decision making is not always rational. We do not always have the time, motivation, or ability to gather and analyze the relevant information, to calculate probabilities, or to consider all of the possible solutions. Instead, we may use heuristics, such as the availability heuristic, when making a decision. The availability heuristic involves basing our judgments about how likely it is an event will occur on the ease with which it comes to mind. Frequent depictions of terrorist events lead the public to overestimate the probability of their occurrence. For example, people worry more about being killed in a terrorist act than in an automobile accident, even though the latter is more likely (Hertwig et al., 2004; Myers, 2001). The reason people are likely to overestimate the likelihood occurrence of an improbable event is due in part to the infrequency of these events. These events receive high media coverage because they are not only infrequent, but also because they tend to be vivid and dramatic (Reber, 2004; Slovic, Fischhoff, & Lichtenstein, 1982). When identifying terrorist threats, in addition to the availability heuristic, we may also be influenced by the representativeness heuristic. The representativeness heuristic involves basing our estimation of the likelihood occurrence of an event on how similar it is to the typical prototype of that event (Kahneman & Tversky, 1982). In terms of identifying terrorists, if one's prototype of a terrorist is someone who is a single, adult, male Muslim, then one is more likely to identify a man fitting this description as a terrorist than as someone who is harmless. The pitfall of the representativeness heuristic is that we often ignore base rate information. Statistically speaking, how likely is it that someone fitting the previous description is a terrorist? Conversely, based on an examination of terrorists, what would constitute an accurate description of a likely terrorist? If the media consistently portrays individuals from a certain ethnic or religious group as terrorists, the public will overestimate the likelihood that they are in fact terrorists. Our reliance on these heuristics—along with the resulting errors in judgment—has serious implications for the policies enacted. As stated above, in our discussion of the illusion of control, these errors may lead us to create policies and practices that fail to address real security threats because we are concentrating our efforts on identifying false positives.

Consistent with this notion, Costanza, Kilburn, and Helms point out that the general public seems content to ignore domestic sources of terrorism arising from militia and survivalist groups, and to concentrate almost exclusively on foreign sources of terrorism. In fact, both the public and the government seem to ignore or even support the efforts of groups such as the Minutemen, who operate along the U.S. border to "protect" citizens from the threat of terrorism. Why are we willing to compromise our constitutional liberties for the perception of security? In essence, we have made foreigners, as well as both illegal and recent legal immigrants, our scapegoats. These groups, because they are readily identifiable and are often the target of existing prejudices, serve as a useful target of blame. In contrast, if we were to look within our own group, long-standing Americans, it would not only be more difficult to identify terrorists but also would create a threat to our cherished group identity.

The counterterrorism and fear chapters help us to understand both public and political response to terrorism and the resulting counterterrorist policies enacted. Now we have to use our understanding and knowledge to educate the public and make better decisions. First, we have to recognize that terrorism can originate from both within as well as from outside our borders—and, as discussed by Borgeson and Valeri, may even include linkages between these two groups. As presented in this book, terrorism and counterterrorism are complex issues. To fully understand and address issues relating to terrorism requires an interdisciplinary approach. We need to draw on the fields of psychology, sociology, economics, political science, and criminology to understand and counter terrorism.

Finally, eliminating the sources of terrorism requires more than an academic understanding. The social psychological theories of terrorism, especially terror management theory, suggest that some negative event exacerbates the situation between two groups, creating an "us versus them" scenario. Differences that were once overlooked or viewed as harmless are demonized and provide the justification for aggression. The negative event is something that threatens the well-being or welfare of one group. Terror management theory explicitly states that factors that make death salient—for example, famine, war, and disease—make terrorism more likely. From a social justice, political, or economic perspective, long-term measures to end terrorism require worldwide efforts to eliminate poverty, inequality, and injustice.

REFERENCES

Affleck, G., Tennen, H., Pfeiffer, C., & Fifield, C. (1987). "Appraisals of control and predictability in adapting to a chronic disease." *Journal of personality and social psychology* 53: 273–279.

Biner, P. M., Angle, S. T., Park, J. H., Mellinger, A. E., & Barber, B. C. (1995). "Need state and the illusion of control." *Personality and social psychology bulletin* 21: 899–907.

Brehm, S. S. & Brehm, J. W. (1981). *Psychological reactance.* New York: Academic Press.

Colvin, C. R. & Block, J. (1994). "Do positive illusions foster mental health? An examination of the Taylor and Brown formulation." *Psychological bulletin* 116: 3–20.

Greenberg, J., Solomon, S., & Pyszczynski, T. (1997). "Terror management theory of self-esteem and cultural worldviews: Empirical assessments and conceptual refinements." In M. Zanna (Ed.), *Advances in experimental social psychology* (Vol. 29, p. 61–139). Orlando, FL: Academic Press.

Greenberg, J., Solomon, S., Pyszczynski, T., Rosenblatt, A., Burling, J., Lyon, D., et al. (1992). "Assessing the terror management analysis of self-esteem: Converging evidence of an anxiety buffering function." *Journal of personality and social psychology* 63: 913–922.

Helgeson, V. S. (1992). "Moderators of the relation between perceived control and adjustment to chronic illness." *Journal of personality and social psychology* 63: 656–666.

Hertwig, R., Barron, G., Weber, E. U., & Erev, I. (2004). "Decisions from experience and the effects of rare events in risky choice." *Psychological science* 15: 534–539.

Kahneman, D. & Tversky, A. (1982). "Subjective probability: A judgment of representativeness." In D. Kahneman, P. Slovic, & A. Tversky (Eds.), *Judgment under uncertainty: Heuristics and biases* (p. 32–47). Cambridge, UK: Cambridge University Press.

Langer, E. J. & Rodin, J. (1976). "The effects of choice and enhanced personal responsibility for the aged: A field experiment in an institutional setting." *Journal of personality and social psychology* 34: 191–198.

Myers, D. G. (2001). "Do we fear the right things?" *Observer [association for psychological science]* 6: 10–19.

Pyszczynski, T., Greenberg, J., Solomon, S., Arndt, J., & Schimel, J. (2004). "Why do people need self-esteem? A theoretical and empirical review." *Psychological bulletin* 130: 435–468.

Pyszczynski, T., Solomon, S., & Greenberg, J. (2002). *In the wake of 9/11: The psychology of terror.* Washington, DC: American Psychological Association.

Reber, R. (2004). "Availability." In F. P. Rudiger (Ed.), *Cognitive illusions.* New York: Psychology Press.

Rodin, J. (1986). "Aging and health: Effects of the sense of control." *Science* 233: 1271–1276.

Rodin, J. & Langer, E. J. (1977). "Long-term effects of a control-relevant intervention with the institutionalized aged." *Journal of personality and social psychology* 35: 897–902.

Schultz, R. (1976). "Effects of control and predictability on the physical and psychological well-being of the institutionalized aged." *Journal of personality and social psychology* 33: 563–573.

Slovic, P., Fischhoff, B., & Lichtenstein, S. (1982). "Facts versus fears: Understanding perceived risk." In D. Kahneman, P. Slovic, & A. Tversky (Eds.), *Judgment under uncertainty: Heuristics and biases* (p. 463–491). Cambridge, UK: Cambridge University Press.

Solomon, S., Greenberg, J., & Pyszczynski, T. (1991). "A terror management theory of social behavior: The psychological functions of self-esteem and cultural worldviews." In M. Zanna (Ed.), *Advances in experimental social psychology* (Vol. 24, p. 91–159). Orlando, FL: Academic Press.

Struch, N. & Schwartz, S. H. (1989). "Intergoup aggression: Its predictors and distinctions from in-group bias." *Journal of personality and social psychology* 56: 364–373.

Tajfel, H. & Turner, J. C. (1979). "An integrative theory of intergroup conflict." In W. G. Austin & S. Worchel (Eds.), *The social psychology of intergroup relations* (p. 33–47). Monterey, CA: Brooks/Cole.

Tajfel, H. & Turner, J. C. (1986). "The social identity theory of intergroup behavior." In S. Worchel & W. G. Austin (Eds.), *Psychology of intergroup relations* (p. 7–24). Chicago: Nelson-Hall.

Taylor, D. M. (1997). "The quest for collective identity: The plight of disadvantaged ethnic minorities." *Canadian psychology* 38(3): 174–190.

Taylor, D. M. (2002). *The quest for identity: From minority groups to generation Xers.* Westport, CT: Praeger.

Taylor, D. M. & Louis, W. (2005). "Terrorism and the quest for identity." In F. M. Moghaddam & A. J. Marsella (Eds.), *Understanding terrorism* (p. 169–186). Washington, DC: American Psychological Association.

Taylor, S. E., Kemeny, M. E., Reed, G. M., Bower, J. E., & Gruenwald, T. L. (2000). "Psychological resources, positive illusions, and health." *American psychologist* 55: 99–109.

Taylor, S. E., Lichtman, R. R., & Wood, J. V. (1984). "Attributions, beliefs about control, and adjustment to breast cancer." *Journal of personality and social psychology* 84: 165–176.

Thompson, S. (1999). "Illusion of control: How we overestimate our personal influence." *Current directions in psychology* 8: 187–190.

Thompson, S. C., Sobolow-Shubin, A., Galbraith, M. E., Schwankovsky, L., & Cruzen, D. (1993). "Maintaining perceptions of control: Finding control in low-control circumstances." *Journal of personality and social psychology* 64: 293–304.

Twenge, J. M., Zhang, L., & Im, C. (2004). "It's beyond my control: A cross-temporal meta-analysis of increasing externality in locus of control 1960–2002." *Personality and social psychology review* 8: 308–319.

INDEX

D

E

F

G

H

I

J

R

S

X

Z